CLEAR AND CONCISE COMMUNICATIONS FOR SCIENTISTS AND ENGINEERS

JAMES G. SPEIGHT

CRC Press
Taylor & Francis Group
Boca Raton   London   New York

CRC Press is an imprint of the
Taylor & Francis Group, an **informa** business

CRC Press
Taylor & Francis Group
6000 Broken Sound Parkway NW, Suite 300
Boca Raton, FL 33487-2742

© 2012 by Taylor & Francis Group, LLC
CRC Press is an imprint of Taylor & Francis Group, an Informa business

No claim to original U.S. Government works

Printed in the United States of America on acid-free paper
Version Date: 20120123

International Standard Book Number: 978-1-4398-5479-2 (Paperback)

Visit the Taylor & Francis Web site at
http://www.taylorandfrancis.com

and the CRC Press Web site at
http://www.crcpress.com

Contents

Preface

Writing and verbally presenting data are the most important means for communicating scientific and engineering work and allow readers to appreciate and evaluate the work of other scientists and engineers.

Technical writing, as practiced by scientists and engineers, must emerge from organized thinking processes that, when converted to writing, illustrate the thoughts of the writers. Scientific and engineering writing must be based on fact and cannot be based on emotions. In fact, scientific and engineering writing is goal-directed and is guided by the writer's goals, which must involve a sense of purpose that has been defined by the writer.

No single course of action can prepare the scientist or engineer for every communication situation that he or she will face. Nevertheless, he or she should be able to handle most situations if there is a preliminary consideration of limitations. One of these constraints is format, and it is necessary to understand that there is no universal format for scientific and engineering writing; the formats used in one organization are not the same formats that scientists and engineers use in a different organization.

In addition, the document must appeal to the designated audience, and the writer must ensure that the tone is appropriate for the readership. The writer must be respectful and polite to his or her readers, and there must be sufficient information about the problem under investigation. Indeed, there must be enough information in the document for the reader to understand the context of the problem.

The way in which many scientists and engineers choose to define a research problem can vary greatly from writer to writer. An important goal for research then will be to discover how this process of representing the problem works and how it affects the writer's performance.

The purpose of this book is to assist the scientist and engineer at every level of education to be able to present his or her ideas in a clear, understandable, and logical manner. As a result, the audience will be sufficiently informed and be able to discuss the ideas at the forum in which the presentation is made or be sufficiently informed by reading the journal article or company report in which the ideas and concepts are published.

The book describes the means to juggle the demands of writing a scientific or engineering paper or report or making a presentation. The book also assists the novice (and perhaps even the mature) scientific and engineering writer to accomplish his or her writing goals.

Finally, my thanks go to my many colleagues who have expressed to me the need and potential for this book. Thanks also go to Dr. Reza Mohammed, formerly of the University of Trinidad and Tobago—currently with RMIT University, Melbourne, Australia—whose talk "Preparing and Delivering an Excellent Presentation" while he was at the University of Trinidad and Tobago provided me with the stimulus and forethought to write this book. Also, my sincerest thanks go to Zaleena K. Chin Yuen Kee, BSc, MSc, of the University of Trinidad and Tobago for providing clarity of thought and preliminary editing of the draft manuscript.

1 Scientific and Engineering Writing

1.1 INTRODUCTION

Writing is the most important means for communicating scientific and engineering work. Research and publication complement teaching (Aristotle, 1991). There are many reasons for writing, one of the most important of which is to better appreciate and evaluate the published work of other scientists and engineers.

Careers in science and engineering are largely supported not only on the quality (which should be the main proviso) but also on the number of publications that a researcher (or a team of researchers) offers his or her colleagues. If the publications are high quality, they lead to research funding and employment. If the publications are numerous (but not always high quality) they may also lead to research funding and employment.

To gauge the contribution of a researcher to science or engineering, other researchers may consciously or unconsciously compute the number of worthwhile publications that a colleague has produced in relation to the number of years he or she has published. The greater speed of release for journal articles means that those who wish to influence their field of study need to publish in peer-reviewed journals in order to quickly communicate their research results.

Documenting the results of experiments in the form of technical papers, company reports, and other necessary documents is an important aspect of scientific and engineering research careers (Jameson, 1995; Carraway, 2006). However, two objectives need to be kept in mind: (1) the readership must find the written work worthy of their time and (2) the work should be cited by other workers in that same technical field. This can be accomplished only if the readers sense that the author is writing with purpose and conviction (Bryan, 1993).

Scientific and engineering research is essential for the growth and development of technology. Only through sound scientific and engineering research can new and old ideas get tested, thus guarding against stagnation and dogmatism.

Conveying one's research findings is an exciting moment because it represents the outcome and recognition of an arduous process (Kotur, 2002). Clarity in reporting how the research was conducted and what results were obtained is paramount for both the research community and advancement of technology. Only through clear and thorough writing can scientists and engineers transfer the benefit of the research for mankind (*para homo*). It is through the correctly written article that the scientists and engineers can appreciate the concepts being developed and judge the extent to which results can be applied in their respective settings. The results serve as a basis on which further scientific and engineering research can be planned and implemented.

A successful publishing career means writing for a highly specific scientific or engineering audience and it can take most scientists and engineers many years to discover how write papers in a manner that results in a high rate of paper acceptance—in fact, some scientists and engineers never achieve the goal of a high acceptance rate and may become doomed to mediocrity.

To acquire competent writing skills, the scientist or engineer can work alone, in isolation from his or her colleagues, and rely on learning from rejection letters and from harsh peer reviews. Alternatively, the novice scientist or engineer can associate with fellow scientists or engineers who will give constructive comments as well as offer support when they read draft manuscripts. If these scientists and engineers are authors in their own right, the novice scientist or engineer can learn much from them by their recommendations for getting a manuscript published.

Thus, writing a scientific or engineering manuscript/report/document is one of the most important tasks facing the scientist or engineer and (to many) may also be one of the most daunting (Cetin and Hackan, 2005). In spite of the fears that can accompany such a task, the satisfaction of seeing one's work survive the peer review process and appear at some future time as a printed article may be immeasurable. On the other hand, the frustration and annoyance involved with the publication process may seem to be an insurmountable hurdle, thereby forcing valuable work to fall by the wayside mainly because the peer reviewers were unable to understand the *raison d'être* and results of the work.

Technical writing, as practiced by scientists and engineers, must emerge from organized thinking processes which, when converted to writing, illustrate the thoughts of the writers. Scientific and engineering writing must be based on fact and cannot be based on emotions. In fact, scientific and engineering writing is goal-directed and is guided by the writer's goals, which must involve a sense of purpose that has been defined by the writer.

Generally, the purpose of scientific and engineering writing is in the publication of data, which marks the endpoint of research that has been performed, completed, peer-reviewed, accepted, and complements teaching. Writing has numerous benefits, one of the most important ones being the inherent training undertaken to better appreciate and evaluate the published work of others. Effective scientific writing is an important component of all scientists and engineers and should be cultivated at an early stage of their respective careers (Hayes and Flower, 1980; Paradis and Zimmerman, 1997; Flower, 2000; Peh, 2007: Dawson and Gregory, 2009).

A successful publishing career means writing for a highly specific scientific or engineering audience and it takes most authors years to discover how to do this in a way that results in a high percentage of accepted papers. An early decision is whether to work alone or with colleagues. To acquire these skills the scientist or engineer can work alone, in isolation from colleagues, and hope to learn from rejection letters and from harsh peer reviews, or he or she can build an informal team of fellow scientists and engineers who are both critical and supportive and who will read and comment on any written work. This is often a quicker, more efficient, and more stimulating path.

If the scientist or engineer is new to an organization or academic department, it is necessary that he or she wishes to quickly determine who will be supportive of his or her aims versus who may be less than helpful. The coffee klatsch group may be of little value but a novice author can learn much from established authors by passing them drafts for their assessment and their recommendations for getting published.

There is a tradition in science and engineering that sees the writing process as a series of decisions and choices but this tradition can only be asserted if the writer is prepared to answer a number of questions about the work once it is published (as for a scientific or engineering paper) or read (as of a project report).

To most scientists and engineers it may seem reasonable to suppose that all of the thoughts are logical, but it is not always clear to the fledgling scientist or engineer how the thoughts might interact, especially in a coordinated manner. The best way to move ahead with writing is for the would-be writer to understand the nature of the thoughts (and choices) that make some forms of writing good and other forms of not-so-good writing, but the ultimate thought is the difference between the two forms of writing.

The need to disseminate scientific and engineering knowledge and expertise for the betterment of humanity led to the concept of writing in scientific journals. Scientific publications that resulted there from provided additional benefits such as reputation among peers and monetary benefits, beyond spreading knowledge. Interestingly, publishing scientific content in journals regularly is a prerequisite for appointment or promotion in several institutions across the globe. Thus, with scientific publications becoming synonymous with job survival, scientists and engineers have started publishing aggressively in recent years.

Scientific and engineering writing in English can be traced back to the fourteenth century when the late Middle Ages in England saw an outburst of scientific writing in the vernacular that moved English discourse in new directions and established new textual genres (Taavitsainen and Pahta, 2004).

Approximately 200 years later The Royal Society (London) established good practice for scientific writing by emphasizing the importance of plain and accurate description rather than rhetorical flourishes as well as the importance of not boring the reader with a dull, flat style (Harmon and Gross, 2007). Despite vast technological advances, there is no reason to expect that scientific and engineering writing is any less important today (Matthews et al., 2000).

In fact, since that time, publication of research data in the form of research papers has come to be regarded as an international currency that transcends political borders. For young scientists and engineers with published articles in internationally reputable journals, they are a great help when applying for positions in foreign institutions and for overseas fellowships.

For more established scientists and engineers, publications enable them to gain recognition and acknowledgment as experts in a particular field at national and international levels. Invitations to lecture at scientific meetings and refresher courses, and appointments as consultants to external agencies, panels of experts, advisory boards, reviewer boards, and editorial boards, are among the benefits of this enhanced professional reputation (Peh, 2007).

Much has been written about the positive and negative experiences acquired from getting scientists and engineers to write clearly and succinctly. In fact, there is a commonly held assumption that increased writing in the science or engineering classroom will automatically lead to improved writing and comprehension by students. This is not always the case because faculty members fail to understand the tenets, strengths, or limitations, nor do they grasp the way in which these features affect writing-to-learn in science. In fact, learning-by-writing occurs only when students know how to use writing-to-learn (Moore, 1991).

One of the critical aspects of the scientific and engineering process is the reporting of new results in scientific journals in order to disseminate that information to the larger community of scientists. Communication of data contributes to the pool of knowledge within a scientific or engineering discipline and very often provides information that helps other scientists and engineers interpret their own experimental results. Most journals accept papers for publication only after peer review by scientists and/or engineers who work in the same field and who recommend the paper be published (usually with some revision) or rejected (for various reasons).

While a request for revision of a work might seem to be inappropriate and bothersome to the writer/author, revision is not to be judged as a last-ditch repair effort but is a constant process of reevaluation of the means by which the technical data are presented. This will lead to a final product that often offers a better understanding of the work and the results because, by the time the revisions are made, the ideas expressed in the document have been allowed to incubate and mentally (with one author) or verbally (with several coauthors) transfer to finalization.

However, the art of communication is not as simple as it might appear. For a scientist or engineer to be able to say or write exactly what is on his or her mind is more than a major hurdle and requires considerable thought and effort as well as a passion and conviction for what is being written or spoken. It should always be kept in mind that even though the facts have been clearly communicated, this does not mean that the individual reader or the collective reader (the audience) will necessarily regard them as important.

Because most scientific journals accept manuscripts only in English, an entire industry has developed to help nonnative English-speaking authors improve their texts before submission. It is just now becoming an accepted practice to utilize the benefits of these services. This is making it easier for scientists to focus on their research and still get published in top journals.

Scientific and engineering writing is a form of communication and is a style of writing used in fields as diverse as gene divergence and process engineering. Scientists and engineers explain their technology and related ideas to technical and nontechnical audiences. Speaking, as well as writing, is also important in science and engineering. Throughout the career of a scientist or engineer, he or she will confront many writing situations, including proposals, formal reports, and journal articles.

In all situations, the scientist or engineer may have to write or present as part of a team consisting of other scientists and engineers, some of whom may be from a completely different subdiscipline of science or engineering. Although collaboration on a document or presentation presents a challenge to the team members, it also has advantages. One advantage is that working as part of a team broadens the range of ideas that the document or presentation can incorporate. Another advantage is that collaborative teamwork allows the group to draw from the various writing and editing strengths of the individual members. In a successful team effort, a strategy develops that emphasizes the advantages and mitigates the disadvantages.

No single course of action can prepare the scientist or engineer for every communication situation that he or she will face. Nevertheless, he or she should be able to handle most situations if there is a preliminary consideration of any. One of these constraints is format (Chapter 6) and it is necessary to understand that there is no universal format for scientific and engineering writing. The format used in one organization is not necessarily the same format that scientists and engineers use in a different organization.

Anyone who has read scientific and/or engineering papers in a journal will have noticed that a standard format is frequently used. This format allows (and encourages) a researcher to present his or her information clearly and concisely.

However, all scientific and engineering writing cannot be treated in the same way and is dependent upon the discipline or subdiscipline of the writer. For example, it is to be expected that chemical thermodynamic writing or reactor engineering writing will be different from one another. In fact, writing is often looked upon as an art form rather than a science. The process of learning to write effectively does not end with any particular guidelines and it continues throughout scientific and engineering careers. Scientists and engineers become apprentices of an art form for which there is no maestro and this describes the writing that scientists and engineers and are required to do.

Regardless of whether a scientist or engineer is writing a technical paper, a slide presentation, an e-mail, or a memorandum, the written communication must represent the abilities as well as the character of the writer (Medawar, 1979; Day, 1994; Markel, 1996; Alley, 1996; Alley et al., 2006). Using an appropriate tone is essential and the language must be considered very carefully so that the writer does not come across as arrogant, overconfident, or too demanding. In all correspondence, the writer should focus on being concise and accurate. Key points should be presented early in all documents so that those same points stand out from the rest of the text.

In addition, the document must appeal to the designated audience and the writer must ensure that the tone is appropriate for the readership. The writer must be respectful and polite to his or her readers and there must be sufficient information for the problem under investigation. Indeed, there must be enough information in the document for the reader to understand the context of the problem.

Addressing such issues will ensure that the written communication helps the scientist and engineer to build and maintain a professional relationship with his or her colleagues and peers.

Scientists and engineers often think of planning as the act of rationalizing or figuring out how to get from here to there and the ensuing formulation of a detailed plan. However, in the planning process writers must form an internal representation of the knowledge that will be used in writing, which is likely to be more abstract than the initial representation will eventually be (Flower and Hayes, 1981).

For example, a whole network of ideas might be represented by a single key word. Furthermore, this representation of the knowledge of the scientific or engineering writer will not necessarily be made in language that is understandable to the reader. Planning, or the act of building this internal representation, involves a number of subprocesses. The most obvious is the act of generating ideas, which includes retrieving relevant information from long-term memory.

Sometimes this information is so well developed and organized in memory that the writer is essentially generating standard written English. At other times the writer may generate only fragmentary, unconnected, even contradictory thoughts, like the pieces of a jigsaw puzzle that has not yet taken shape. When the structure of ideas already in the writer's memory is not adequately adapted to the current writing task, the subprocess of organizing takes on the job of helping the writer make meaning, that is, give a meaningful structure to his or her ideas. The process of organizing appears to play an important part in creative thinking and discovery since it is capable of grouping ideas and forming new concepts.

More specifically, the organizing process allows the writer to identify categories, to search for subordinate ideas that develop a current topic, and to search for superordinate ideas that include or subsume the current topic. At another level the process of organizing also attends to more strictly textual decisions about the presentation and ordering of the text. That is, writers identify first or last topics, important ideas, and presentation patterns.

As scientists and engineers proceed with their writing task, a new element enters the task environment, which places even more constraints upon what the writer can include. Just as a title constrains the content of a paper and a topic sentence shapes the options of a paragraph, each word in the growing text determines and limits the choices of what can come next. Yet, the influence that the growing text exerts on the composing process can vary greatly.

When the writing is incoherent, the text may have exerted too little influence; the writer may have failed to consolidate new ideas with earlier statements. On the other hand, one of the earmarks of a basic writer is a dogged concern with extending the previous sentence and a reluctance to jump from local, text-bound planning to more global decisions, such as the nature of the work to be covered.

The continuing task of writing a paper or a report usually makes large demands on the writer's time and attention. However, in doing so, it is competing with two other forces that could and also should direct the composing process; namely, the writer's knowledge stored in long-term memory and the writer's plans for dealing with the problem. It is easy, for example, to imagine a conflict between what the writer knows about a topic and what he or she might actually want to say to a given reader.

Part of the drama of writing is seeing how scientific and engineering writers integrate the multiple constraints of their knowledge, their plans, and their text into the production of each new sentence and thus into the work as a whole.

Finally, before a scientist or engineer commits words to paper in a scientific or engineering manuscript, he or she should understand the subject matter that is being communicated to the reader or the audience. In other words, pages should not be filled with (often meaningless) sentences unless there is a general (even fixed) idea where those sentences are headed. Even after there is a general understanding of the subject matter, the scientist or engineer should not begin writing until the various writing constraints (which are those aspects of the writing that the writer does not control) have been analyzed. These constraints include (1) the audience for the document and (2) the format of the document. Another constraint, not discussed here, is mechanics, which comprises grammar, punctuation, and usage (Chapter 5 and Chapter 6). Besides the constraints of audience and format, there is also the writing style (Strunk, 1918; Strunk and White, 1959) to be considered (Chapter 5) but this is an aspect of the writing that the scientist or engineer does control.

Finally, many manuscripts submitted for publication in journals contain unnecessary technical terminology, unmanageable descriptions of the work that has been done, and convoluted figure legends. Prior to submittal, the manuscript must be edited and revised by the author(s) so that it is grammatically correct, logical, clear, concise, and uses consistent terminology, which is consistent with that used in previous papers published in the journal. This process is enhanced if the authors have written the manuscript in a simple and accessible style, which is the best way to convey the message of the paper and to persuade readers that it is important enough to spend time reading.

In addition, many journal editors know that the impact of a paper is maximized if it is as short as is consistent with providing a focused message. Authors can place all the technical information (figures, protocols, methods, tables, additional data) necessary to support their conclusion as Supplementary Information (SI), which is published only to accompany the published print/online paper. A Supplementary Information section is peer-reviewed, and the use of such a section emphasizes the impact of the conclusions, which is enhanced by being presented in concise and focused form in the print/online journal. It is not surprising that many editors will encourage authors to use a Supplementary Information section to enhance the impact of the print/online version, especially when providing the full details of methods and protocols in the print/online version may mask the message.

1.2 DEFINITIONS

Throughout this book certain terms are used interchangeably to indicate scientific and engineering writing. It is opportune at this time to present the terms and their respective definitions as employed herein.

The term *scientist* or *engineer* refers to a person who wishes to pursue science or engineering as a career whether the degree be a baccalaureate degree, a master's degree, or a more advanced degree or qualification (such as a doctor of philosophy degree) in one of the scientific or engineering disciplines. In one way or another, a scientist is a person who has scientific training or who works in the sciences. On the other hand, an engineer is someone who is trained to work as an engineer.

Furthermore, *science* is regarded as inclusive of the life sciences, physical sciences, mathematics, and social sciences (the scientific study of human society and social relationships). *Engineering* is viewed as a field that includes all specialties such as civil, mechanical, electrical, petroleum, and computer engineering. However, in all cases, the system of education of scientists and engineers should be organized around realistic experience.

The practical difference between the scientist and the engineer lies in the educational degree and the description of the task being performed by the scientist or engineer. It is widely believed that scientists explore the natural world and discover new knowledge, while engineers apply that knowledge to solve practical problems, often with an eye toward optimizing cost, efficiency, or some other parameters.

A *technical paper* (be it scientific or engineering in nature and content) (Chapter 2) is the means by which scientific and engineering ideas and work are broadcast to a readership of peers. The paper will be published in a journal of some repute, that deals with the subject matter of the paper and also has as wide a circulation as possible.

A *technical report* (again, be it scientific or engineering in nature and content) (Chapter 2) is a document forwarded to a funding agency or to company management in which the status of a project is presented to the agency or to management.

On the other hand, a *progress report* (Chapter 11) is written to inform a supervisor, associate, or customer about progress that has between made on a project over a certain period of time. The necessary constituents of a progress report are included elsewhere (Chapter 11) rather than here because the progress report can take any one of several forms. The report may appear as (1) a letter, (2) a memorandum, (3) a fax, or (4) an e-mail, depending on the request of the funding agency or the project sponsor. The format of the progress report will be made clear in the contract or, if not, by the funding agency or the sponsor.

A *presentation* (Chapter 9) is a verbal project update using the medium of slides (commonly Power Point slides) on a particular (usually research) project. The presentation may be made to a peer group at a local, regional, national, or international scientific or engineering meeting. The presentation may also be made to a committee established by a funding agency or to company management. Other forms of communication also exist (Chapter 11).

Following from the above, the terms *scientific writing* or *engineering writing* as used in this text do not refer to the genre of *science writing* or *technical writing,* which can take the form of newspaper and magazine articles, memoirs, books, correspondence, presentations, instructions, or blogs. This kind of writing has different goals and intended audiences than writings of the scientific and/or engineering community. Popular science and engineering writers must report, describe, *and* communicate clearly (Ferris, 2001). Science and engineering writers address the larger public about the science and technology that shape modern life, as well as the broader social issues that are influenced by science and engineering.

1.3 CAREER WRITING

The fundamental purpose of scientific and engineering writing is not the mere presentation of information and thought, but rather the communication of the information to the audience. It does not matter how pleased an author might be to have converted all of his or her data into sentences and paragraphs; it matters only whether a large majority of the reading audience accurately perceives what the author had in mind.

Moreover, the appearance in print of several good papers by a scientific or engineering author is an enhancement to his or her career. Similarly, the appearance within a company of several meaningful project reports and/or presentations by a company scientist or engineer author can mean upward mobility to the author.

The phenomenal increase of both primary papers and review articles is placing scientists and engineers under increasing pressure to keep up with the literature in their respective fields of interest. Furthermore, as the performance of text retrieval and analysis algorithms to draw meaningful information from the literature improves, scientists and engineers will increasingly rely on these to harvest relevant papers from the deluge of available information.

However, scientists and engineers will still read papers if they think that the title is interesting or that the message or question being answered is important. And the better the paper is written and the more logical its arguments, the higher the chances that the reader will proceed beyond the abstract and find it convincing enough to cite. Consequently, it is of utmost importance to keep two things in mind throughout the writing process: the main message and the reader. The author's goal must be to convince the reader that this is important research. If a paper ignores readers' interests, they in turn might ignore the paper. This book will offer help in this respect.

Indeed, scientists and engineers should not complain that there is a lack of guidance—there is an abundance of literature on how to write clearly and understandably to attract the interest of the readers (Strunk and White, 1959; Zinsser, 1976). These and other books like them are unknown to, or ignored by, most scientists. Although such books might not cater explicitly to scientific writing, they are nevertheless valuable as they explain how to organize material in a coherent way, and how to write a manuscript that is both informative and readable.

More important, such books convey an important message: authors should write not for themselves but for their readers. Many scientists would do well to heed this advice, as a clear and understandable manuscript is more likely not only to draw citations but also to be accepted for publication in the first place. Unfortunately, the scientific and medical literature is still abundant with lengthy, unclear prose that is likely to confuse readers, even those who are familiar with the subject material.

There are limitations on the style and format of a scientific manuscript. In addition to taking into account the specific requirements of scientific journals, a paper must generally have an introduction, separate sections on methods and results, and a discussion of the results in relation to the original hypothesis. The very nature of a scientific paper—presenting and discussing results in an unbiased way—also poses restrictions on the writing style: the passive voice is ubiquitous in order to appear impersonal, and the need to cite relevant references can interrupt the concise and clear flow of text. However, these rules are flexible enough to allow a paper to be written in both an informative and interesting way. It is essential to understand the basic rules on how to present results in a way that is more likely to attract interested readers.

It must always be remembered by the writer that the purpose of writing a scientific or engineering research paper is not only to present results but also to explain, interpret, predict, suggest, hypothesize, and even speculate. The main purpose of the discussion is to provide a forum in which the author seeks to convince the reader of the logical experimental setup, the soundness of the results, and the validity of the speculations.

At every step, it should be clear to the reader whether the discussion merely interprets results and predicts further outcomes, or launches into more far-fetched speculations. References are essential for this process, but readers are easily annoyed if they are dragged through every publication that has a bearing on the main theme. For the most part, readers expect a coherent interpretation of the results and a demonstration of their relevance.

If the discussion must include intellectual or literary mumbo jumbo, the results are obviously not sufficiently convincing on their own. If reviewers and editors feel this way, they might require additional experiments before accepting the paper.

Thus, the process of scientific or engineering writing is best understood as a set of distinctive thinking processes that scientists and engineers orchestrate or organize during the act of composing the written work. Moreover, these processes have a hierarchical, highly embedded organization in which any given process can be embedded within any other. The act of composing a technical work is, in itself, a goal-directed thinking process, guided by the writer. Furthermore, scientists and engineers create their own goals in two key ways: (1) by generating both high-level goals and supporting subgoals that embody the writer's developing sense of purpose, and (2) by changing, when necessary, the major goals or even establishing entirely new goals based on what has been learned in the act of writing.

Not surprisingly, science and engineering works are often hard to read. Most readers (including scientists and engineers) assume that the difficulties of reading technical works are born out of necessity, out of the extreme complexity of scientific concepts, data, and analysis. This is (or should be) far from the truth.

Scientists and engineers do not simply read; they interpret. The old adage *three attorneys in a room means nine different opinions* is equally applicable (although the formula changes) to any piece of prose, no matter how short, which may be interpreted in ten or more different ways to ten different readers. This is based on the recognition that readers make many of their most important interpretive decisions about the substance of prose based on clues they receive from its structure.

Therefore, in order to understand how best to improve writing, it is necessary for scientific and engineering *authors* of documents to understand better how *readers* read such works. Such an understanding will help to produce a methodology based on the concept of reader expectations. This is often experienced through the format of a scientific or engineering document.

The format of scientific and engineering documents may seem confusing for the beginning scientific or engineering writer due to its rigid structure, which is very different from writing in other disciplines.

One reason for using this format is that it is a means of efficiently communicating scientific findings to the broad community of scientists in a uniform manner. Another reason, perhaps more important than the first, is that this format allows the paper to be read at several different levels. For example, many people skim *titles* to find out what information is available on a subject. Others may read only titles and Abstracts. The realization here must be that the scientific or engineering format helps to insure that at whatever level a person reads the paper (beyond title skimming), the reader will likely get the key results and conclusions.

However, the complexity of the subject should not lead to mental impenetrability of the text or the presentation. There are a number of rhetorical principles that can produce clarity in communication without oversimplifying scientific or engineering issues or even fostering misunderstanding of the writing. Indeed, improving the quality of writing and presentation improves the quality of thought.

The fundamental purpose of scientific and engineering discourse is not the mere presentation of information and thought, but rather the actual means of communication. It matters little how pleased an author might be to have converted all of his or her data into sentences and paragraphs. What really matters is whether or not a majority of the reading audience accurately perceives what the author was attempting to communicate.

Therefore, in order to understand how scientists and engineers can improve their writing and communication skills, it is necessary to understand the skills and habits of the readers. Such an understanding will help to produce a writing and communication methodology based on the concept of reader expectations.

It is also necessary to recognize that correct English usage (English or American English) is critical in science writing. In fact, scientists and engineers often try to be so concise that the English usage should be better than that of workers in other disciplines. Furthermore, if English is not the first language of the writer, proofreading by a native-speaker might be helpful.

Finally, being able to write well is a real challenge, in part, because it is difficult for many scientists and engineers to explain exactly what *good writing* is. Good writing of any sort has some general characteristics, or traits, that make it good:

- It has a clear theme,
- There is background information defining the context of the theme, and
- There are illustrative examples supporting the theme and any conclusions.

Good scientific and engineering writing, however, often has another important trait—the author uses quantitative data and graphics to illustrate and support the theme of the paper. Incorporating quantitative data in a paper is not easy because it requires very precise descriptions that may be difficult to craft.

As with anything else in life, however, a scientist and engineer can become a better writer with practice, both by the act of writing itself and by reading examples of good writing.

Finally, a well-written scientific or engineering document must explain the theory behind the work as well as the experimental design and execution and the meaning of the results. Thus, the document must be written in a style that is clear and concise since the authors' purpose is to inform an audience of other scientists and engineers about an important issue and to document the particular approach they used to investigate that issue.

An outcome of this book is to show the means by which a scientist or engineer can become a writer or build a system that would follow the writing process. But first, it is necessary for the would-be writer to understand the three paths to success:

1. First, the scientist or engineer needs to define the major elements or subprocesses that make up the larger process of writing—such subprocesses would include planning, retrieving information from long-term memory, and reviewing/revising.

2. Second, the scientist or engineer needs to show the manner in which the various elements of the process interact in the total process of writing—for example, whether knowledge about the audience is actually integrated into the moment-to-moment act of composing the work.

3. Finally, since a written work is often used as a primary a tool for thinking, the writer would want the work to support critical questions in the discipline that have led the need for the research described in the paper or report—it may help the writer see things he or she did not see at the time of commencement of the writing.

Thus, at the beginning of composing the work-to-be, the most important element is obviously the rhetorical problem itself.

1.4   PURPOSE OF THE BOOK

Thus, the purpose of this book is to assist the scientist and engineer at every level of education to be able to present his or her ideas in a clear, understandable, and logical manner. As a result, the audience will be sufficiently informed and be able to discuss the ideas at the forum in which the presentation is made or be sufficiently informed by reading the journal article or company report in which the ideas and concepts are published.

Novice writers are often thrown into writing at the deep end of the writer's pool with recommendations such as *take my paper as an example, and do the same* from a well-meaning professor or mentor. Unfortunately, good examples are rare and not identified as *good examples*, and good writing is usually rewritten bad writing. In fact, typically, the submitted paper has rarely been rewritten enough to qualify as a good example—pressure to meet the publication deadlines, lack of time for writing the paper, and many other debatable reasons are involved.

Scientific and engineering writing, a form of technical communication, is a style of writing used in the scientific and engineering fields that is used to explain scientific and engineering technology and related ideas to scientific and engineering and nonscientific and engineering audiences. Writing by a scientist or engineer is a means of gathering and collating information from existing documentation and from subject matter experts.

The reasons for scientific and engineering writing can range from noble to base reasons. Topping the list is altruism, where one writes for the pleasure derived from the creative activity of writing and from sharing one's intellectual pursuits, as well as for the desire to advance human knowledge for the benefit of mankind. For these authors, writing is a channel for expressing the joy of scientific or engineering discovery. At the bottom of the list, writing may be considered by some scientists and engineers to be a chore where getting published is a *necessary evil* in order to fulfill certain specific minimum requirements, e.g., for getting a job or a promotion (Peh, 2007). Another reason is because of *abject laziness*. Members of this last group are often the loudest critics of scientists and engineers who publish research papers at their morning coffee or afternoon tea breaks in the common room or in the staff lounge.

The most compelling reason for many scientists and engineers for writing and publishing their work is to fulfill specific job requirements by employers (such as universities). Some researchers may have a driving force to contribute to advancements in scientific and engineering knowledge or may simply love their work and want to share it with others.

Still other researchers may work in a group or environment where journal articles are essential for professional survival (Carraway, 2009). Such a unit or environment might include initial appointment to an academic position, renewal or confirmation of that appointment, promotion to a higher-level appointment, and granting of tenure. And in this very competitive world where *publish or perish* is the axiom (especially in universities), publication is a necessity (Carraway, 2006).

Whatever the motive, the scientist or engineer will need something important to say if his or her results are to be published.

Communication is the lifeblood of every scientist and engineer; however, the means by which that communication is carried out are varied, and different types of communication are suitable based upon the type of information that needs to be conveyed. One specific area of communication involves transmitting an understanding and knowledge of a scientific or engineering subject to others that need to know the basics and details of the subject in question. In written form, this type of communication is called technical writing or a scientific or engineering publication. In the visual form, this type of communication is called a scientific or engineering presentation (Alley, 2003).

Scientists and engineers read far more than they write—both read far more than the average reader and what they read is far more complicated. It is often thought (even stated) that it is possible to become a better writer by building on reading skills. However, no matter how much reading the writer has under his or her belt, if the writing is not clear, concise, interesting, fluid, and organized, there will be no enjoyment in the discovery of new information. If the experience of reading a scientific or engineering paper is not pleasant and if what is read is not clear, the author of the paper has failed in his or her mission to add further knowledge to his or her area of expertise.

In fact, having a paper in print means absolutely nothing, other than only one or two reviewers found some indication of possibly new knowledge in what was submitted. If readers do not go to the paper, the writer has accomplished nothing. If other scientists or engineers read the paper and but they remain unconvinced as to its use or at the very least to verify its findings, the writer has accomplished nothing—except for publishing information that adds nothing to the advancement of science or engineering.

The paper will not be cited and, other than a line in the author's résumé, the published paper falls in the deep chasm of meaningless science or engineering. There are not many men (or women) like the Austrian monk Gregor Mendel who, having published the results of his work in an obscure journal, was discovered more than fifty years later and whose work was to become one of the building blocks of the science of modern genetics.

Therefore, two objectives must be clear and these are to (1) attract the reader to the paper, and (2) convince both reviewer and reader of the worth of the scientific or engineering contribution.

The first objective is to get the work approved for publication, and the second objective is to get the author(s) cited as making a meaningful contribution to the technical literature of that discipline. And it is worth remembering that the reviewer/referee whom the journal editor asks to review the paper will give the author(s) two to three hours of time to evaluate the work, but the reader will give the author(s) less than two minutes if the writing is not clear and succinct.

Therefore, a good scientific and engineering writer needs strong language and teaching skills and must understand the many conventions of modern scientific and engineering communications.

For scientific and engineering documents to be useful, readers must be able to understand and employ them without having to decode wordy and ambiguous prose. Good scientific and engineering writing clarifies scientific and engineering jargon; that is, it presents useful information that is clear and easy to understand for the intended audience. Poor scientific and engineering writing may increase confusion by creating unnecessary scientific and engineering jargon, or failing to explain unavoidable scientific and engineering terms that a reader would not be expected to be familiar with.

Scientific and engineering writing is a form of communication to convey a particular piece of information to a particular audience for a particular purpose. It is a means of publishing ideas and theories about scientific subjects and scientific and engineering subjects associated with science and engineering.

Scientific and engineering writing translates complex scientific and engineering concepts and instructions into simpler language in order to enable readers and users of the described technology to perform a specific task in a specific way. To present appropriate information, scientists and engineers must understand the audience and its goals—audience analysis is a key feature of all scientific and engineering writing.

Technical writing rarely, if ever, is done in a manner that personalizes it. There are never references to the first person. The structure of the text is always done in a detached or third party reference point. The text is geared to teach information; therefore, the tone is that of a teacher instructing a student in the subject. This at times may appear as simply a list of steps to take to achieve the desired goal, or it may be a short or lengthy explanation of a concept or abstract idea.

In order to write an article that is both logical and convincing, the scientist or engineer must gather the information from other sources, such as technical documents, reports, references, and manuals/handbooks. Upon assembling all the information gained, the writer must gauge the audience and the background that the reader has in the field being written about. Higher levels of education or experience will allow the document to be written referring to concepts, abbreviations, acronyms, and terms commonly used by those that work in that field. For audiences that are not familiar with the field, the writer must do much more explanation and education about the basics. He or she must assume that the audience has no knowledge of the topic about which he or she is writing. Depending on the type of document, these considerations may be more or less important.

Scientific and engineering writing refers to straightforward explanations and/or instructions dealing with a particular technical subject and the writing must be easy to understand and follow. As with any writing project, the level of understanding by the audience must be considered in the development of the document. When writing about technical information, it is important to be concise and exact about the subject matter.

Finally, writers are generally unable to see what in their writing makes readers stumble. Writers typically find (egotistically) that what they write is extremely clear—because they understand it and fail to understand why others do not find it clear. Unfortunately for scientists and engineers of all ages and levels of maturity, the reader is the person who has to reconstruct the writer's thoughts. As a result, writers need the help of their readers to identify what, in their thought-to-word translation process, failed to work as smoothly as expected.

Furthermore, courses related to improvement of writing skills are often considered by scientists and engineers as being appropriate for schoolchildren but not for the super-egoed scientists and engineers. These courses are writer-centric, not reader-centric and focus on sentence structure, grammar, rules, and more rules, and more exceptions to the rules.

Thus, a major goal of this book is to assist the reader in the development of effective scientific or engineering writing skills. Written and oral communications skills are probably the most universally sought qualities by graduate schools and professional organizations. The reader is responsible for developing such skills to a high level.

The book will describe the means to juggle the demands of writing a scientific or engineering paper or report or making a presentation. In practice, scientific and engineering writers frequently emphasize the various constraints of writing as those of an almost insurmountable problem. In fact, redefining the problem in this way is obviously an unacceptable strategy for any scientist or engineer since the new representation may not fit reality. Furthermore, if a writer's representation of his or her rhetorical problem is inaccurate or simply underdeveloped, then he or she is unlikely to solve or attend to the missing aspects of the problem.

In summary, defining the research problem is a major, immutable part of the writing process. But the way in which many scientists and engineers choose to define a research problem can vary greatly from writer to writer. An important goal for research then will be to discover how this process of representing the problem works and how it affects the writer's performance.

In such circumstances, this book will assist the novice (and perhaps even the mature) scientific and engineering writer to accomplish his or her writing and research goals.

REFERENCES

Alley, M. 1996. *The Craft of Scientific Writing*. 3rd edition. Springer-Verlag, New York.

Alley, M. 2003. *The Craft of Scientific Presentations*. Springer-Verlag, New York.

Alley, M., Schreiber, M.M., Ramsdell, K., and Muffo, J. 2006. How The Design of Headlines in Presentation Slides Affects Audience Retention. *Technical Communication*, 53(2): 225–234.

Aristotle. 1991. *The Art of Rhetoric*. H. Lawson-Tancred (Editor). Penguin Books, London, United Kingdom.

Bryan, J.G. 1993. On Scientific Writing: The Need for More Conviction and Subjectivity. *The Leading Edge*, 346–347.

Carraway, L.N. 2006. Improve Scientific Writing and Avoid Perishing. *American Midland Naturalist,* 155(2): 383–394.

Cetin, S., and Hackam, D.J. 2005. An Approach to the Writing of a Scientific Manuscript. *Journal of Surgical Research*, 128: 165–167.

Dawson, N.V., and Gregory, F. 2009. Correspondence and Coherence in Science: A Brief Historical Perspective. *Judgment and Decision Making*, 4(2): 126–133.

Day, R.A. 1994. *How to Write and Publish a Scientific Paper*. 4th edition. Oryx Press, Phoenix, Arizona. Pp. 8–14.

Ferris, T. (Editor). 2001. *The Best American Science Writing*. Harper Collins Publishers, New York.

Flower, L.S., and Hayes, J.R. 1981. A Cognitive Process Theory of Writing. *College Composition and Communication*, 32(4): 365–378.

Flower, L.S. 2000. Writing for an Audience. *Language Awareness: Readings for College Writers*. 8th edition. P. Eschholz, A. Rosa, and V. Clark (Editors). Bedford/St. Martin's, Boston, Massachusetts. Pp. 139–141.

Harmon, J.E., and Gross, A.G. 2007. *The Scientific Literature: A Guided Tour*. University of Chicago Press, Chicago, Illinois.

Hayes, J.R., and Flower, L.S. 1980. Identifying the Organization of Writing Processes. In *Cognitive Processes in Writing: An Interdisciplinary Approach*. Lee Gregg and Erwin Steinberg (Editors). Lawrence Erlbaum Associates, Hillsdale, New Jersey.

Jameson, B. 1995. The Benefits of Publishing Technical Papers. *The Exchange: Newsletter of the Scientific Professional Interest Committee*, Winter: 372–374.

Kotur, P.F. 2002. How to Write a Scientific Article for a Medical Journal? *Indian Journal of Anesthesiology*, 46(1): 21–25.

Markel, M. 1996. *Technical Communication*. 4th edition. St. Martin's Press, New York. Pp. 420–421.

Matthews, J.R., Bowen, J.M., and Matthews, R.W. 2000. *Successful Scientific Writing: A Step-By-Step Guide for the Biological and Medical Sciences*. 2nd edition. Cambridge University Press, Cambridge, United Kingdom.

Medawar, P.B. 1979. *Advice to a Young Scientist*. Harper & Row, New York.

Moore, R. 1991. Does Writing About Science Improve Learning About Science? *Journal of College Science Teaching*, February: 212–217.

Paradis, J.G., and Zimmerman, M.J. 1997. *The MIT Guide to Science and Engineering Communication*. The MIT Press, Cambridge, Massachusetts.

Peh, W.C.G. 2007. Scientific Writing and Publishing: Its Importance to Radiologists. *Biomedical Imaging and Intervention Journal*, 3(3): 55.

Strunk, W. 1918. *Elements of Style*. Humphrey, Geneva, New York.

Strunk, W., Jr., and White, E.B. 1959. *The Elements of Style*. Macmillan Publishers, London, United Kingdom.

Taavitsainen, I., and Pahta, P. 2004. *Medical and Scientific Writing in Late Medieval English*. Cambridge University Press, Cambridge, United Kingdom.

Zinsser W. 1976. *On Writing Well: An Informal Guide to Writing Nonfiction*. Harper & Row, New York.

2 Types of Scientific and Engineering Writing

2.1 TECHNICAL PAPER

The goal of scientific engineering writing is clarity and understanding: the purpose is to communicate specific ideas in clear and understandable writing that follows a set of basic principles (Table 2.1 and Table 2.2), and everything about the document should contribute to this goal (Medawar, 1979; Day, 1994; Alley, 1996).

Some scientists and engineers believe that writing reports, papers, and making presentations are not part of research, but an adjunct to it or even an undesirable distraction. This view is inaccurate.

The purpose of research is to increase the store of human knowledge, and so even the very best work is useless if the scientist or engineer cannot effectively communicate it to the rest of the world. Additionally, even if he or she believes that she understands the ideas and concepts behind the work, he or she is likely to find that when the time comes to write or speak them, he or she is unable to clearly enunciate them.

The process by which an author is able to clarify his or her thinking, of which writing papers and giving talks is one aspect, is a valuable part of improving research.

2.1.1 ORGANIZATION

A scientific or engineering paper should communicate the main ideas of the research (such as the techniques and results) early and clearly. Then, the body of the paper can expand on these points; a reader who understands the structure and big ideas can better appreciate the details. This advice also applies at the level of sections and paragraphs. The writer should not commence with a mass of details, hoping that the reader will somehow manage to discern which of those are relevant to the main point of the paper—which is either omitted or emerges later—like some novels where the murderer appears for the first time in the last chapter.

For each section of the paper, the writer might wish to consider writing a mini-introduction that says what its organization is, what is in each part, and how the parts relate to one another. For the whole paper, this is probably a paragraph. For a section or subsection, it can be as short as a sentence. This may feel redundant to the writer but readers have not spent as much time with the structure as has the writer, and so they will appreciate the mini-introduction as means of orienting themselves within the text.

Some scientists and engineers prefer to write the Abstract, and often also the Introduction, as the final steps. Doing so makes them easier to write, because the rest of the paper is already complete and can be described easily.

TABLE 2.1

Principles of Good Writing

1. Write correct English—do not use jargon.
2. If acronyms are used, spell them out in full when used for the first time.
3. Be consistent with names—refer to each significant character (algorithm, concept, language) using a proper name.
4. Join each subject to a verb that expresses a significant action.
5. In each paragraph, move the reader from familiar information to new information.
6. Be coherent—choose subjects that refer to a consistent set of related concepts.
7. Order the text—the reader should be able to see how concepts relate to one another.
8. Abstract—convey the essential information found in the paper.

TABLE 2.2

Good Writing Practices

1. Practice writing in brief, daily sessions.
2. Focus on the process, not the product; ensure consistency and quality.
3. Prewrite; make notes, diagrams, and anything else that can help.
4. Use index cards or some other form of organization to plan a draft.
5. Initially, do not be concerned about page limits; write the paper and then cut it down to size.

However, there are other scientists and engineers who prefer to write these sections early in the process (and then revise as needed), because they are used to frame the paper.

A scientific or engineering paper need not be written as a chronological narrative of all the things that were tried or tested and it is not necessary to devote space to the paper proportionately to the amount of time spent on each task. The purpose of research invariably includes the exploration of blind alleys and dead-ends to enable the scientist and engineer to do the small amount of work that is worth reporting. In short, the purpose of the paper is not to describe all that has been done, but to inform the reader of the latest results—unless the work done allows the reader to build upon the nonresults.

In the latter case, the discussion should focus on differences from the successful technique, and if at all possible should provide general rules or lessons learned that will help others to avoid such blind alleys in the future.

The scientist or engineer should write for the readers, rather than writing for himself/herself. In particular, the writer should not focus on what he or she personally finds most interesting. It is a very common error for the writer to become immersed in the technical approach or the implementation details without having appropriately framed the problem. The writer should first state the nature of the problem before discussing steps.

The writing should be *to the point* and any text that does not support the point of the work should be eliminated. To do this, the writer needs to examine each section of the paper in turn and ask himself/herself what role it serves and whether it contributes to the main point of the paper. If there is no contribution, the section should be rewritten or deleted.

The writer should examine each paragraph within each section and ask himself/ herself whether that paragraph has a single point, and whether that point contributes to the goals of the section. If not, the paragraph should be rewritten or deleted.

Next, the writer should examine each sentence within each paragraph. If it does not make a single, clear point that strengthens the paragraph, the sentence should be rewritten or deleted.

Finally, within each sentence, each word should be examined and those that do not strengthen the point should be replaced or deleted.

This entire process may need to be repeated several times in order to keep a fresh perspective on the paper.

2.1.2 WRITING STYLE

Passive voice has questionable value in technical writing. It often obscures the point of the work. Use of the active voice with simple, clear, and direct phrasing is preferred.

Use of the *first person* is rarely appropriate in scientific and engineering writing and should never be used to describe the operation of a program or system. It is only appropriate when discussing something that the author of the paper did manually.

The writer should make every word count and if a word does not support the point, the word should be deleted—excess verbiage makes it harder for the reader to appreciate the point of the paper (Strunk, 1918). Short and more direct phrases should be used wherever possible. Self-congratulation and value judgments should be avoided—the writer should present the facts and allow the reader to be the judge.

Words such as *obviously* or *clearly* often offend readers by insulting their intelligence, and by demonstrating the writer's inability to communicate intuitively.

Above all, writing more clearly will help the scientist or engineer to think more clearly and often reveals flaws (or ideas!) that had previously been invisible even to the writer.

Furthermore, if the writing is not good, readers may not be able to comprehend any good ideas presented in the paper or the readers will be suspicious of the experimental work. If the scientist or engineer does not (or cannot) write well, there is no reason for readers to believe that the writer had a coherent research plan and experimental style.

2.1.3 FIGURES

Most scientific and engineering papers use figures (sometimes referred to as *charts*). The old adage *a picture speaks a thousand words* is very true in the area of scientific and engineering writing.

Different people learn in different ways, so a textual presentation should be complemented with visual ones. Visual inclusions offer an alternate presentation of the ideas and concepts and fill gaps or enable the reader to verify his or her understanding of the work (Chapter 9). Figures can also help to pull a skimming reader into the text (or at least communicate a key idea to that reader), and make the paper more visually appealing and understandable.

A figure should stand on its own, containing all the information that is necessary to understand it. Good captions contain multiple sentences; the caption provides context and explanation. The caption may also need to explain the meaning of columns in a table or of symbols in a figure. However, it's even better to put that information in the figure proper; for example, use labels or a legend.

When the body of a document contains information that belongs in a caption, there are several negative effects. The reader is forced to search throughout the document to understand the figure. As a consequence, the flow of the reading and understanding is interrupted with details that are relevant only when the reader is looking at the figure.

2.1.4 NAMING

Each concept in the paper should have a descriptive name—generic terms and acronyms are not reader friendly and suggest to the reader that the writer may not fully understand or even agree with the concept!

Terms should be used consistently and precisely—*elegant variation*, which uses different terms for the same concept, can be confusing to the reader and may not emphasize different aspects of the concept. While elegant variation may be appropriate in some nontechnical writing, it is not acceptable in scientific and engineering technical writing, where the writer should clearly define terms when they are first introduced and then use them consistently. The reader of a scientific or engineering paper expects that use of a different term will indicate a different meaning, leading to confusion, and the reader and may even miss the point if the writer changes wording gratuitously. Choose the best word for the concept, and stay with it. Furthermore, do not use a single term to refer to multiple concepts.

2.1.5 FEEDBACK

The writer should always be ready to seek feedback. It is crucial that the paper is completed well in advance so that the writing can be improved before requesting feedback from one's colleagues.

Even a scientist or engineer rereading his or her own text after being away from it for a time (three weeks is a convenient time period) can find items that were not noticed in the original writing. And if this is the case, one wonders what an outsider might find. Furthermore, when readers misunderstand the paper, the writer must accept some of the blame, even if it is apparent that the reviewers have missed the point; the writer will learn how the work can be misinterpreted, and eliminating those ambiguities will improve the paper.

The golden rule is *do not waste anyone's time if there are major flaws in the work*. Check the work thoroughly to discover any serious problems.

The most efficient way is to get feedback sequentially rather than in parallel. Rather than asking three people to read the same version of the paper, ask one person to read the paper, and then make corrections before asking the next person to read it. This prevents the writer from getting the same comments repeatedly—hopefully the subsequent readers can give the writer new feedback rather than repeating what he or she already knows, and the comments will not be diametrically opposed to the comments of the preceding reviewer(s). Asking multiple reviewers at the same time might be considered to be devaluing their time and indicating that the writer does not mind if the reviewers waste their time.

Finally, the writer should be willing to reciprocate when colleagues need comments on their papers. By helping colleagues, the scientist and engineer will not only help them but also learn what to emulate or avoid, and the colleagues will be more willing to review the writer's draft papers.

2.1.6 MISCELLANEOUS

The writer should be consistent such as using a consistent number of digits of precision—one decimal place or two decimal places or three decimal places, but not all three in the same paper. The journal for which the paper is destined may offer guidelines on the number of decimal places required for precision.

In addition, the writer should bear in mind the message he or she wishes to convey to readers—too many digits of precision can distract readers from the larger trends and the big picture. Indeed, including an inappropriate number of digits of precision can cast suspicion on all of the results, by giving readers the impression that the writer is statistically naive.

A related work section should not only explain what research others have done, but in each case should compare and contrast that to the work in the paper. Additionally, for each significant piece of related work, after reading related work, readers should understand the key idea and contribution of that work.

2.1.7 REJECTION

Scientists and engineers who submit technical papers to journals will experience rejection. In some cases, rejection indicates that the author(s) should move on and begin a different line of research but in most cases, the comment offered by the reviewers offer an opportunity to improve the work. This brings an opportunity to craft a good paper ro appear at a later date rather than have a poor paper appear earlier.

The wrong lesson to learn from rejection is discouragement or a sense of personal failure. Many papers—even from authors who later win awards—have been the objects of rejection (at least once) and as the writer returns to the work to make recommended changes, the work product will improve.

It is to be remembered that receiving feedback on a paper will help the writer to improve it. On the other hand, the scientist or engineer does not want to get a reputation for submitting poor-quality work. Here is a general rule of thumb: if the writer knows the flaws that will make the referees/reviewers reject the paper or the valid criticisms that they will raise, do not waste their time submitting the paper. A writer should only submit a paper if he or she is not aware of flaws or errors of logic.

2.2 PROJECT REPORT

A *project report* is a means that a scientist or engineer can use to communicate to a client (or funding organization) the findings of a project, thereby enabling the scientist or engineer to keep the funding organization or the sponsor informed of the progress of the project and to raise any items for their attention (Chapter 11) (Sandia, 1990). The project can, in the current context, be any scientific or engineering study or research of a problem or question, or the gathering of information on a technical subject.

The project report also advises other members of the project team whether or not the project is on track and likely to finish within schedule—the timeliness of the project should be evident to other members of the project team from the regular project meetings and the report merely reduces the meeting conversations to writing.

On the other hand, a *progress report* (Chapter 11) is written to inform a supervisor, associate, or customer about progress that has between made on a project over a certain period of time—usually a shorter period (a month)—where a project report is typically written annually.

The necessary constituents of a *progress report* are included elsewhere (Chapter 11) rather than here because the progress report can take any one of several forms. The report may appear as (1) a letter, (2) a memorandum, (3) a fax, or (4) an e-mail, depending on the request of the funding organization or the project sponsor. The format of the progress report will be made clear in the contract or, if not, by the funding organization or the sponsor.

The production of good technical writing for a project report (or any technical report) is as much a part of the project as doing the experimental work. However excellent and original a piece of work the project may be, unless the results can be communicated to other people (of varying disciplines outside of the scientific and engineering disciplines) it may be of no value. Communicating results of an investigation in a clear and useful way is a key part of science and is the reason for devoting a lot of effort to this aspect.

While the format of the report will vary with the dictates of the organization or the audience for which the report is written, following the format presented in the previous section may suffice in the absence of any distinct or prescribed formatting guidelines. Generally, the subsections of a scientific or engineering report will include: (1) Title and authors, (2) Abstract, (3) Table of Contents, (4) Introduction, (5) Experimental techniques and methods, (6) Results and discussion, (7) Summary/ Conclusions, (8) References, and (9) Appendices (if necessary).

2.2.1 FOCUS OF THE REPORT

Project reports often have to be written for a varied readership—for instance, technical managers and financial managers. Thus, the use of appendices and summaries can help the writer be understood by a wider readership than is the case for a scientific or engineering paper. In fact, the range of the readership will determine not only the approach but also the technical level and the style of the writing.

Thus, the objectives of a report must appeal to the needs of the reader, the information involved, and the appropriate tone, within the clearly defined limits of the report specification. The report must always identify the objectives of the project, the information that is covered in the report, and for whom it is written—when the objectives are clearly identified, the actual writing of the report will become much easier.

A project report (or a progress report; see Chapter 11) is used to communicate all aspects of the status of the project, usually to coworkers, designated colleagues for review, and especially management. The various aspects that need to be included in such a report are as follows:

- Technical progress: any issues that have been identified and their effect on the project
- Schedule: whether or not the project is on time
- Budget: the expenditures and whether or not they are within the budget
- Staffing: the manpower effort that has been expended to date
- Deliverables: what has been delivered and whether or not they have met targets
- Risks: any risks that have been identified and their effect on the project

Thus, with the project report, the project manager can clearly document the overall status of the project and keep the client regularly informed of the project's progress. In addition, any issues related to risks can be raised and advice given to the client. The project team is also kept well informed (no one should be able to claim "I did not know") of the progress made to date and the report also serves to ensure that all project successes are clearly communicated.

Writing a project report can, for a variety of reasons, be dependent on the nature of the project. Using a structured approach to writing the report will help to produce a clear and more effective report. There are also two common project reports: (1) status reports and (2) summary reports that are produced during project execution.

Thus, a good report is easy to recognize—the title is precise and informative, its layout and format are well organized, and the binding is easy to handle and opens flat to reveal both text and diagrams. Reading a well-written report can be a pleasure: the style is accurate, fluent, and concise, with headings to indicate the content of each section. There are no absolute rules about the details of report production, because every report must be totally adapted to the needs of its reader.

In keeping with writing a scientific or engineering paper as described in the previous section, a project report must also focus on the subject and the reader. In fact, the reader is the most important person. In this respect, the report should be checked thoroughly for technical errors, typing errors, and inconsistency.

The report may vary in length depending upon the subject matter and (assuming the report is a company document) may be as short or as long as the company dictates. In fact, the report should be organized for the convenience of the report user and the writing should be accurate and to the point. The references should be correct in all details and the right figures/diagrams/charts and the tables with correct and explanatory captions (that can act as stand-alone items if transplanted into the presentation) should be in the right place for the convenience of the reader.

The summary (usually an Executive Summary, since the report may be for management) should give the whole picture, in miniature—this is in contrast to an Abstract, which may be brief and used to whet the appetite of the reader for the contents of the document.

2.2.2 PROJECT STATUS REPORT

A *Project Status Report* or a *Project Summary Report* are types of a *Project Report* used by teams to communicate the status of a project (see also Chapter 11). The Project Report is distributed regularly and is used to inform the client whether the project is on track. The Project Report specifies whether the project is ahead of schedule, under budget, and adequately resourced, and it is one of the key tools used to ensure success.

Project Reports are used when the project manager (the lead scientist or engineer) needs to communicate the status of a project. These might be typically completed on a monthly (or weekly) basis (depending on the requests from the client as presented in the project contract) and ensure that they are sent to the client (as well as others he or she may designate).

Therefore, the information in a Project Report or a Project Status Report must be accurate and up-to-date and the report must only include information that is relevant to the client. The scientist or engineer writing the report must ensure that all aspects of the project status are covered in the report. The client may have provided a project report template, which, if this is the case, should be followed carefully.

Completing a regular Project Status Report is a core project management activity. It allows the project scientist or engineer to keep staff on track, keep management informed of the progress of the project, and seek guidance when needed, especially if unforeseen issues arise (Chapter 11). It also helps the project manager to maintain the entire project team as a cohesive unit and focuses the team on delivering the project on time and within budget.

Thus, the production of a good piece of technical writing for a project report is as much a part of the project as doing the experimental work for the project. However excellent and original a piece of work the project may be, unless the results can be communicated to other people, it may as well not have been done!

Communicating results of an investigation in a clear and useful way is a key part of science and is the reason for devoting a lot of effort to this aspect.

The main part of the report should be comprehensible to the client. If more detailed information is to be included about some aspects (for instance, a complicated mathematical derivation, of which only the result is essential to the main discussion), consider including this as an appendix.

Unless advised otherwise by the client, it is not necessary, or even desirable, to describe every minute detail of what was done. One of the most important aspects of good technical writing is to be concise, yet remain informative. The ability to select what is essential, and to omit what is merely incidental detail, is a skill every scientist needs to develop.

In view of this, the main part of the report must be within the word limit(s) specified in the applicable project description or project contract. An overlong report may be penalized by the client refusing to read such detail and/or sending it back for rewriting.

Project reports can cover a host of potential subjects such as product evaluation, feasibility study or performance assessment. The content must be planned and structured.

The essential purpose of content planning is to create a logical structure to clearly present the key messages of the project report. Through well-organized presentation of the key messages and their supporting ideas, the writer is able to either inform or influence the reader as appropriate. If this planning step is skipped, the content is likely to lose its impact as the ideas will not be clearly expressed and as easily understood.

When the ideas and planning stages have been thought through and structured, the report writing can begin. The focus should involve writing the following:

- An *Introduction* to provide context and background.
- The *Presentation* of key findings (sometimes called the *Results and Discussion* section of the report) and supporting points according to project structure and logical order.
- The *Conclusions* or *Executive Summary* that the client is more likely to read first—the Executive Summary typically is placed at the front of the report before the Introduction.

2.2.3 EXECUTIVE SUMMARY

An *Executive Summary* is a term used for a short document that summarizes a longer report, a proposal, or a group of related reports in such a way that readers can rapidly become acquainted with the subject matter without having to read the whole document.

The Executive Summary will usually contain a brief statement of the problem or proposal covered in the major document(s), background information, concise analysis, and main conclusions. It is intended as an aid to decision making by business managers and will differ from an *Abstract*, which is usually shorter and is intended to provide an overview rather than being a condensed version of the full document. Abstracts are extensively used in scientific and engineering research publications where an Executive Summary would be superfluous.

The Executive Summary typically completes the report (either a full project report or even a project status report) and is often the part of the report that is read first. In fact, not only managers and executives but many technical readers may read the executive summaries only and may not even consider reading the remainder of the report.

Executive Summaries go by many different names—sometimes the executive summary is called an Abstract, the usual designation in scientific and engineering papers. However, Abstracts differ from Executive Summaries, because an Abstract is typically written for a scientific or an engineering purpose. Generally, an Abstract consists of 250 words or less. On the other hand, the Executive Summary can be as long as several pages and will appear after the transmittal memo and just before the first page of the report.

Other characteristics of the executive summary are variable since many organizations and businesses have document template definitions to suit their own purposes. An executive summary will typically (1) be written in nontechnical language, (2) consist of short and concise paragraphs, (3) have a conclusion, and (4) make a recommendation.

Thus, in the Executive Summary the writer will place the issue under investigation (the *problem*) and the *purpose* for the work in the first paragraph. The *scope and limitations* of the work as well as the *procedures* will be placed in the next paragraphs. The *analysis and decisions* will comprise the final paragraphs.

2.3 FUNDING PROPOSAL

In the early days of scientific and engineering research and before the Industrial Revolution, most scientific and technological research was carried out by individual scientists or engineers using their own funds; although the cost of the search for the Philosopher's Stone was often sponsored by kings and other members of the nobility who sought an easy route to gold.

After the Industrial Revolution, scientific and technological research became increasingly systematized as companies discovered that continuous investment in scientific and engineering research and development was the key to success in business competition. As a result, research was encouraged and funding for research became more formal and competitive, resulting in the need for written proposals or formal presentations to the funding organization.

A funding proposal is usually written in response to a *request for a proposal* (RFP), which is an invitation for scientists and engineers to submit a proposal for a specific project. The process of issuing a *request for a proposal* brings order to a funding (sometimes in the government called *procurement*) and allows the program of work to be identified clearly.

The *request for a proposal* may be issued by a government agency as a general announcement of program direction, or it may be issued by a company to one or two specific groups of scientists and engineers within a university. The organization releasing the request for a proposal will usually dictate to varying degrees the exact structure and format of the response. Effective *requests for a proposal* typically reflect the strategy and short-term/long-term business objectives of the organization issuing the *request*. The *request* should also provide detailed insight upon which the responder will be able to offer a detailed, but relevant, work program.

On the other hand, a government department of a private organization may fund research through grants to academic scientists and engineers or to other researchers outside of academia. The rule for the response to receiving such a grant may be the same as the rule for responding to a *request for a proposal*. In either case, a proposal will need to be written and submitted to the grant-awarding organization.

The primary rules of proposal writing are (1) Murphy's law—if anything can go wrong it will go wrong; (2) Speight's 1st Law—Murphy was an optimist; and (3) Speight's 2nd Law—do not get the reviewers annoyed and/or frustrated.

A proposal seeking research funding is also, to the dismay of some scientists and engineers, a scientific or engineering document. It must therefore be technically sound but there are other attributes that the organization providing the funds for the project may require as part of the proposal.

In all cases, the author(s) of the proposal must check to determine if the organization has any specifications for the document format. The answer is inevitably *yes*. The parts considered necessary are described below insofar as every request for a proposal (RFP) seen by the author contained these requirements for the proposal.

Finally, it must be recognized by the scientist or engineer that the results of sponsored research are the property of the sponsor (government or private). The government may allow publication as soon as the data are made available through reports whereas the private organization generally will not allow such release of data until patent protection has been sought.

It is also necessary to realize that funding of research by private companies is mainly motivated by the *good of the company* and that private companies are much less likely than government agencies to fund research projects solely for the sake of knowledge. The *good of the company* incentive will cause researchers to concentrate their energies on projects that are perceived as likely to generate profits.

Consequently, it is likely that the project-generated data will remain the in-house intellectual property of the company and will never be released for publication in the open literature. In addition, government-sponsored research can result in collaborative projects between several organizations whereas privately funded (company-funded) research remains localized to one researcher or within one group of researchers.

In any case, it is more than likely that the researcher(s) will be required to sign a nondisclosure agreement.

2.3.1 PROJECT TITLE/COVER PAGE

The title should be unambiguous and should present to the reader/reviewer the major focus of the proposal/project. It is preferable to use a single sentence for the title. If the sentence is long and cannot be shortened, there is the possibility that a two-part title will be acceptable. The title should be an Abstract of the entire proposal.

If the proposal is a response to a Request for a Proposal (RFP), it may be appropriate to use the title presented with the *request for a proposal* as the title of the proposal.

Usually the Title/Cover Page includes signatures of key people in the proposing organization. If the work will involve collaboration with another organization the name of the participants and the organization should also be included on the title page.

In terms of document presentation, the cover should look professional and neat and does not need an artistic report cover or any other procedure that may send the funding organization an uncomfortable message (about how the funds will be spent) on unnecessary artistic endeavors.

2.3.2 PROJECT OVERVIEW

Many authors consider the Project Overview as an Executive Summary; however, he or she should rest assured that many agencies insist that the report be read in detail and the busy reviewer who only has enough time to read the Project Overview will have little influence on the final outcome (i.e., whether or not to fund the proposal). However, details that are clarified at a later point in the proposal should not be included in the Project Overview.

The Project Overview should give each reviewer a feeling for the subject of the proposal and how the work will assist in the presentation of new knowledge on that subject.

The author may also wish to use the Project Overview to show that he or she (the team) has knowledge of the organization from which the research funds are requested. In addition, if collaboration with another organization is planned, the author should ensure that the interests of the collaborating organization are also included.

Many authors prefer (some will actually insist) that the most appropriate time to prepare the Project Overview is after the proposal has been completed and all aspects of the proposal are clear and understandable. The Project Overview will, without doubt, form a strong impression in the mind of the reviewer and ensure that his or her questions will not start at this point. A reviewer should complete his or her reading of the Project Overview by nodding his or her head in agreement with the description contained therein.

2.3.3 BACKGROUND INFORMATION/STATEMENT OF THE PROBLEM

Many authors consider this section as a review of *relevant literature* in which previous work involving similar studies is cited. This will show the reviewer (and thus, the funding organization) that the author(s) are aware of (and familiar with) other work that has preceded the proposal.

This section should be readable and understandable, and the language should minimize the use of jargon, buzzwords, acronyms, and the like. Trying to be one of the *in-crowd* (*one of the boys* or *one of the girls*) by the use of such words will not work with most reviewers. Nothing frustrates a reviewer more than having to continually translate jargon, buzzwords, and acronyms.

By the use of plain and readable language, the author should place his or her project proposal in relation to other efforts and show how the project: (1) will extend the work that has been previously done, (2) will avoid the mistakes and/or errors that have been previously made, (3) will serve to develop stronger collaboration between existing initiatives, or (4) will be sufficiently unique insofar as the project will not follow paths successfully explored previously by other researchers.

The author(s) should also use the statement of the problem to show that the proposed project is definitely needed and should be funded. Name-dropping of previous workers is not always advantageous at this point. For example, in the area of petroleum science and technology, name-dropping of prior authors can be the literal kiss-of-death for a proposal. Much of the work published in some areas of these fields is a mere copy of prior work performed decades ago (without reference to the prior work) and is wrong/incorrect/poorly-planned/useless for advancing technology.

Thus, it is essential that in any proposal it should be clearly indicated how the goals and objectives of the *request for a proposal* will be addressed in the proposal, and how the proposed work plan will help the funding organization in fulfilling its own goals and objectives.

2.3.4 PROJECT DETAIL

2.3.4.1 Goals and Objectives

The author(s) should differentiate clearly between the *goals* and *objectives* of the work.

The *goals* are the statements of what the work will accomplish, which are usually difficult to measure quantitatively but do create the setting for the proposed work.

On the other hand, the *objectives* are operational, indicate the specific accomplishments of the project, and are therefore measurable. The objectives form the basis for the activities and, hence, the evaluation of the project. In this respect, measurable objectives for the project should be presented in the proposal. Funding agencies love to be presented with a checklist to determine if the work is making any headway or merely spending money.

2.3.4.2   Experimental Methods

There should be a very clear link between the experimental methods described in this section and the objectives defined in the previous section. The writing should be explicit (*to the point*) with statements showing how the experimental methods will satisfy the objectives of the project and contribute to delivering answers to problem(s) on which the proposal is focused.

There should also be statements showing what is new, unique, or innovative and the author(s) of the proposal should ensure that any innovative aspects are clearly presented and obvious to the reviewer and/or the organization.

Also, when collaborative relationships are involved for the project, there should be descriptions of the work to be developed with any collaborative groups. A good way to show collaboration is in the experimental methods described in the proposal and in which of these methods will be carried out by the collaborators.

This section should clearly indicate how the experimental methods that will be used will allow the outcomes of the project to have value for others beyond the project.

2.3.4.3   Staff/Administration

This section is typically used to describe the roles of the different people associated with the project and the importance of each person or group. The author should make sure to clarify how each of the roles are essential to the success of the project and how each role clearly relates to the methods.

In this section, key persons should be named along with (1) title, (2) experience, and (3) qualifications. The descriptions of the personnel should inform the funding organization that excellent people are involved in, and committed to, the project. Teamwork is often preferred by funding agencies.

There may also be the need for an Advisory Committee (Steering Committee) to assist in the project. If details of the Committee are not requested elsewhere in the RFP, this is as good a place to describe the Committee, how it will be organized, and the function of the Committee.

2.3.4.4   Available Resources and Needed Resources

Most funding agencies prefer to see collaborative ventures involved in a proposal. This allows the organization to determine the means by which the funds are brought together with other existing organizations that are involved in dealing with the needs and objectives of the organization.

It is advantageous for a prospective funding organization to be advised that local resources have already been contacted and there are plans to include them in the project. Indeed, letters from local resources supporting the project are an excellent addition to the proposal.

Needed resources refer to members of the staff/administration section showing the personnel to be paid from the project funds. They should be identified by name along with short descriptions of each person who will be involved in the project and supported by the funding.

The descriptions should clarify in the mind of the potential funding organization that these people are ideally suited to conduct the project. And, instead of having all full-time staff on the project, a number of part-time staff may satisfy the needs of the project.

2.3.4.5   Facilities

Though not always necessary or requested, it may be helpful to provide a brief description of the facilities that will be used for the project. Existing facilities can be described and used as *in-kind contributions* for the project. This will show the funding organization the amount of additional money that would be needed if the facilities were not donated.

2.3.4.6   Equipment/Supplies/Communication

It is wise to list the equipment that will be used for the project with brief descriptions of how the equipment will be used. However, requesting purchase of new equipment is another matter.

Funding sources are usually less willing to provide funds for equipment purchase than they are for the support of personnel involved in the project.

If new equipment is required, there had better be a good reason and the cost must have been thoroughly researched for inclusion in the budget.

However, including the cost of a new coffee machine to go with the morning donut break for the team members is *not* a wise inclusion in the budget and is more than likely to suffer immediate rejection as well as have an undesirable influence on the reviewers and the funding organization.

2.3.4.7   Budget

Following on from the comments made in the last section, the budget *must* be realistic. The author needs to carefully determine exactly what funds will be needed from the funding organization to carry out the project.

An inflated budget will be obvious to the reviewer(s) and the budget should be reviewed by someone in the proposing organization for realism. In addition, budgets are rarely spent equally in each year of the project. There is the need to determine if a large amount of funding is required at the beginning of the project or if the activity of the project will be increased over a period of time. It is not very realistic to expect a new project to be able to be up and operating (with the accompanying expenditure of considerable sums of money) during the first 6 months or year of operation.

A good strategy to use with a potential funding organization is to request a small amount of funding for the first year of a multiyear project, with a description of the anticipated achievements with the minimal funding during that year.

At this time, it would be appropriate to describe (unless the funding organization states otherwise in the RFP) the budget categories that will be used for reporting to the organization. However, it is very likely that these categories will be specified by the organization. If they are not, the author may wish (informally) to ask for guidance on this issue.

2.3.4.8   Evaluation Plan

It is important (if nor essential) to describe in the proposal the means by which the work will be measured against the objective.

An Evaluation Plan will show to the prospective funding organization the success of the project and, moreover, that the organization's investment in the project was worthwhile. The Plan may take the form of a checklist for comparison against the objective specified in the proposal or a questionnaire. The Evaluation Plan does not have to be excruciatingly complex (to the point of pain and frustration) but it is important to indicate to the prospective funding organization that the author has not forgotten this important step.

A worthwhile Evaluation Plan should also include some sense of concern for what goes on following the conclusion of the funding period relating to (1) the means by which the initiatives that have been started under the project be sustained, and (2) the future of any new and innovative discoveries made during the work.

2.3.4.9   Appendices

Appendices are typically included to illustrate those aspects of the project that are of secondary interest to the reviewer and the organization. The Appendices will provide additional information for the reviewer and the funding organization.

An Appendix might also include a clear indication (through a graphical representation—a Gantt chart) of the *time line* for the project and the times when each aspect of the project will be implemented.

Letters of support are also often included in a separate Appendix. This will let the funding organization know that the project has the support of other persons of some consequence. The letters must be substantive and addressed directly to the funding organization.

2.4   REPORTS TO A COURT

An *expert report* to a court is a report written (usually) by one expert that states findings and may offer opinions. On some occasions, the report may have multiple authors. This decision is made by the court or by the attorneys representing either side of the case.

An expert report is generated by an *expert witness* and will offer an opinion on points of controversy in a legal case. The expert witness is typically sponsored by one side or the other in a litigation case to support that party's claims, but he or she (the expert witness) should not take sides but *report the truth to the court*. The report should state facts, discuss details, explain reasoning, and justify the conclusions and opinions of the expert witness (Speight, 2009).

In science and engineering, the report written by an expert witness is a critical assessment of a scientific or engineering topic—for example, an independent assessment of the use of a particular type of process or reactor in a petroleum refinery.

Reports to a court by an expert witness (Expert Witness Report), usually by a scientist or engineer whether oral or written, can range from minimal to extensive. Moreover, the scientist or engineer prepares a technical report that is to be read and assimilated (for the most part) by nontechnical personal, other than the expert witnesses hired by opposing counsel (Speight, 2009).

Many attorneys request technical reports other than opinion reports. This aspect depends on the needs of the attorney and must be discussed very early in the arrangement. By no means should the expert witness ever send anything in writing until a request is made by the attorney.

In fact, it is best at first that communication between the expert and attorney occur only in person or by telephone. Written material (including any fax or e-mail) sent to the attorney may not be designated by the court as confidential and is, therefore, discoverable. All documents and exhibits or material pertinent to the case can be examined by the opposing counsel at the time of discovery. Some courts have ruled that even the notes made by the expert witness that are relevant to the case may be given to the opposing counsel during the discovery process.

The expert witness must always keep in mind that all of his or her reports and written communications may be shown to the opposing counsel.

2.4.1 ORAL REPORT

An *oral report* is, of course, a word-of-mouth report that is made by the expert to the judge. The judge may request by-attendance-in-court oral reports from each of the experts involved in a case, which may then take the form of a visual presentation (Chapter 10). Alternatively, although less likely, the judge may request oral reports from the experts by telephone, allowing each expert to speak on different occasions and blind to the other expert reports. Many judges maintain contact with the attorney and opposing counsel and may request oral reports from them on a semi-regular basis.

However, for the expert witness, oral reports should not be a once-during-a-project event but should occur regularly insofar as the expert needs to keep the attorney up to date. Such reports can be made by telephone or in person.

As part of the report to the attorney, the expert informs the attorney as to what has been done, and the conclusions. If the expert has formed an opinion, the attorney should be advised of this opinion. He or she may or may not be looking for certain things from that particular aspect of the evidence, so it is important to work with the attorney in doing that.

One of the field most neglected by expert witnesses is that of presentation techniques. Bad presentations are also one of the most damaging, discouraging, and dispiriting aspects of any case. Presentations may be so bad that a certain winner can turn out to be a loser.

For the presentation, the expert should be aware of the situation. The courtroom situation can be intimidating and dramatic. The expert should be prepared to respond adequately, in which case reading from a script by the expert (no matter how well the script is prepared) is not recommended!

The expert should believe in what he or she is doing. A visibly detached attitude—reading a script as if it did not belong to, or had not been prepared by, the expert—may well be a source of irritation for the judge and the jury, if the jury is present.

In the United States, the language of the court is English—if English is not the mother (or father) tongue of the expert, a typewritten, double-spaced, font-size-12 copy of what the author is going to say should be prepared. This does not mean that the expert should read everything from the paper, but he or she should have it there and ready in case of necessity.

The expert should determine beforehand (from the attorney) the amount of time that is allotted for the presentation. The judge may give a time limit of, say, twelve to fifteen minutes for the expert to include all of the salient facts. The expert should adhere to the time limit. As a general rule, one letter-size page, double-spaced, twelve-point font is equivalent to approximately three minutes of speaking time. The expert should read the text aloud at home, and note the time taken. The speaking style should be slow and clear and allowance made for pauses.

It is advisable to use and prepare media other than the expert's own voice. But this should be integrated into the presentation. In this sense, the expert should make certain that the necessary technical equipment is there in the courtroom (or be ready for the judge's chambers) in time, and in working order. If the expert is to use a blackboard, make sure it is clean and, if possible, have some of what is going to be used on the blackboard already written out before the start of the presentation. If the presentation is to take place in the judge's chambers, an oral presentation only without equipment may be the easiest to do.

The most important part of any presentation in the expert's oral presentation is his or her voice. Everything else should be auxiliary to the expert's role and function in the presentation.

A good presentation should allow for a discussion by or questions from the judge. The presentation is for the judge and she should be treated with the courtesies that that would be given to the Chief Executive Officer of any major company.

As far as mimics, gestures, and positioning as well as movement in the courtroom are concerned, there is no one form that fits all. Some experts will feel (and look, and come across) better sitting at a desk while speaking—being seated is at the discretion of the judge. Some experts use gestures very sparingly while others will gesticulate agitatedly while pacing to and fro across the room, enacting their presentation to the hilt, and still not look ridiculous.

Ultimately, the expert's performance during the presentation is an issue of personal taste, but the expert should make sure that it is his or her own choice of style. The positioning and style of presentation is a conscious decision, based on the expert's own abilities and the demands of the situation. Most of all, it must be the expert's own style; the presentation must come across convincingly.

An expert report is a document that will provide the judge with a written handout that he or she can use for reference. Of course there should be an evident relation between what the expert is saying and what is on the handout. Ideally, the handout (1) summarizes and highlights the main points of the presentation, (2) presents, in written form, citations of the sources, (3) may provide additional information that is not part of the presentation, and (4) if it is to be used, should be in the hands of the judge *before* the expert starts speaking!

2.4.2 WRITTEN REPORT

Written reports form the basis for pretrial preparation, settlement negotiations, and testimony during trial. They may lead to a decision not to call the expert witness, or a settlement, and hence prevent a trial. Reports may be a few paragraphs or voluminous. If a written report is requested, incorporate only what is necessary. Gratuitous and unimportant comments are to be avoided. If a report requires permission, consent, or a waiver of confidentiality, insist that counsel get the proper authorization. The expert must be aware of the due date for the report and to whom it is directed.

If a *written report* is requested, the expert witness should be sure to get guidelines from the attorney who will ensure that there is no conflict with any existing court-directed formatting or guidelines. Furthermore, the report should list the facts as the expert witness knows them and any information obtained from other sources should be referenced.

If the report is related to the opinion, above all, the opinion expressed must be clearly stated to be the expert witness's own opinion. The attorney may try to influence the expert witness in some phase of the report to enhance the attorney's position and court strategy. This must be resisted at all costs!

An expert's report is one of the most important services provided to retaining counsel. A well-written report is immensely helpful to retaining counsel and may well lead to future referrals and the ability to charge premium fees. A poorly written report can and will be used to impeach the expert in the case at hand and future cases for years to come.

Written reports often ensure greater accuracy than oral reports and can serve as a basis for drafting individualized interview questions. However, if this part of the process is confidential, written reports may also increase the chance that investigative information will find its way into the public arena. When a judge feels strongly that confidentiality cannot be maintained once reports are reduced to writing, the judge may want to rely on an oral report (see above).

For a written investigative report, the report should indicate the reliability of the information and the expert must be aware of the scrutiny a written report will undergo.

The data and other information relied on as the basis for opinions and conclusions must be identified. As in any technical document, credibility comes from using appropriate sources of information (Chapter 3). These sources, whenever possible, should have first-hand information. If secondary sources are used, they should be identified as such. Copyright restrictions must be considered when using information published by someone other than the report author. Exhibits used to support opinions and conclusions must be succinctly identified and appended to the report. How the exhibits are used and reported is determined by the writer for overall clarity and their contribution to understanding the report.

The authenticity and validity of the report must be acknowledged by the signature of the writer. Original signatures are preferred; facsimile signatures and signature stamps should not be used. An otherwise professional report with something other than an original author's signature dramatically reduces the professionalism of the report.

2.5   THESES

A thesis is required for an advanced degree and is typically in part fulfillment for that degree. On occasion a thesis is required for a baccalaureate degree and requires the same attention that an advanced thesis requires.

In some ways, writing a thesis is no different than writing other scientific or engineering papers, and much of the advice that appears elsewhere in this book will be relevant to the thesis writer. However, there are some important differences between writing a thesis and writing a technical paper.

Most theses relate to topics that are almost always initially too big and try to include too much. The student/candidate for the scientific or engineering degree should begin his or her preliminary writing as the time passes so that he or she will be well prepared for the final task related to the thesis. In addition, many scientific and engineering students are truly on their own and are left to plan the work as it unfolds—a mentor may be of so little value that he or she is not just useless but advertently or inadvertently places obstacles in the way of the students.

Thus starting thesis planning early allows the student to make a thorough review of the work and make adjustments or shifts in the plan path where necessary. The student will continue to find pertinent material throughout the entire time he or she is working on the thesis.

In addition, for any thesis that depends on physical research and analysis (i.e., the typical scientific and engineering thesis), the student thoroughly discusses the timetable with his or her advisor (assuming that the advisor is not one of the useless ones). Issues that the student may have not considered (e.g., equipment availability) may be out of his or her control and will have significant influence on the timetable.

As to the actual writing of the thesis: While most thesis writers are writing as they read (at least to take notes or to write short summaries of existing scholarship), the majority of the writing is accomplished during the winter term.

2.5.1 READING STRATEGIES

It is important for the scientific or engineering student/candidate to understand and accept that he or she is not going to know exactly what he or she is looking for in the beginning. Initially, the student should read to explore, and will find that certain aspects of the topic interest him/her more than others, and that certain approaches offer more opportunities for new scholarly work. Even if he or she is doing scientific or engineering experimentation (many scientists and engineers may be involved in paper studies), he or she needs to be flexible in the beginning and willing to modify the initial question. It is quite understandable that as a program evolves, the questions at the start can be refined and others arise.

2.5.2 SIMULTANEOUS WRITING AND RESEARCH

The most useful approach is for the student to *write all through the research process*. This includes (1) taking notes while reading and (2) writing summaries or short treatises for everything that is read. It is also a good idea to keep a journal (with at least one or two backup copies as some of these notes and summaries can be incorporate into the thesis and thus, there will be a record of the sources.

In short, the research process should not be viewed as separate from the writing process, which is usually performed at the very end of the thesis timetable.

Commencing work on a draft before the end of the preliminary research is a worth-while effort.

There will be many changes to make to the writing as the project proceeds but the writing process itself often serves to assist in answering some of the questions leading to the need for more research. Most ideas do not coalesce just by reading and need the accompanying writing—writing throughout the research process keeps the thought process active and records the writer's responses to new and different ideas as they are hatched.

2.5.3   Making a Start

Before they begin to research and to write, many scientific and engineering students think of a thesis as just a really extended technical paper. But while the size of the project is overwhelming, the nature of the thesis is actually more complex than a matter of size. There is little, if any, comparison between even a research paper and a thesis.

Indeed, because of the size of the work, breaking the thesis up into smaller components of things to do and items to discuss is the easiest way to make the project manageable.

The *to do* plan is a list of tasks, due dates, and literature that must be read. The *to discuss* plan is the list of the points to be made and how they can be made in the most logical manner. If either of these plans appears to be unwieldy, each chapter can be considered to be a scientific or engineering paper with its own distinct argument. But the chapters must be connected by good, strong transition paragraphs, and each chapter must contribute clearly and coherently to the larger argument.

2.5.4   Thesis Structure

When considering a structure for the thesis, be sure to look for and consider relationships between ideas. These connections should be noted and an attempt on the first outline made as soon as there is the notion of the vaguest form of an organizational outline. The outline will change as the thinking evolves, but each outline that is created will be helpful in keeping track of the evolution of the ideas and in determining the shape of the argument that is eventually used.

Once the outline is formed, it will be easier to think in terms of chapters rather than in terms of the thesis as a whole. It may be that chapters are good units to try to research, write, and edit one at a time. However, it is important that the author allows significant time in the writing process to synthesize these smaller units into a unified and coherent document.

2.5.5   Revision Process

Most students write *at least* two full drafts of a thesis, as well as numerous complex revisions of problem sections and individual chapters. Here is a list of questions to be used during the revision process to ensure that the product of the process is thorough and effective:

- Do the argument and purpose remain clear throughout the thesis?
- Does the introduction to the thesis clearly introduce the idea behind the work and explain the significance?
- Does the body of the thesis present the major points in a logical order?
- Is each of the major points supported by the appropriate amount of evidence and analysis?
- Does the writing make clear transitions from point to point?
- Does the conclusion follow logically from the introduction and body of the work?

It is at this point that a reliable supervisor/advisor/mentor is crucial to the revision process. He or she should be able to detect any problems in the argument as well any other scholar in the field. In fact, other professors and other thesis-writers in the department may be very helpful.

2.5.6 LITERATURE REVIEW

Conducting a literature review is a means of demonstrating an author's knowledge about a particular field of study, including vocabulary, theories, key variables and phenomena, and its methods and history. Conducting a literature review also informs the student of the influential researchers and research groups in the field. Finally, with some modification, the literature review is a *legitimate and publishable scholarly document*.

Apart from the above reasons for writing a review (i.e., proof of knowledge, a publishable document, and the identification of a research family), the scientific reasons for conducting a literature review are many. Prominent among these is that the literature review plays a role in: (1) delimiting the research problem, (2) seeking new lines of inquiry, (3) avoiding fruitless approaches, (4) identifying recommendations for further research, and (5) seeking support for grounded theory (Gall et al., 1996, Hart 1998).

Another purpose for writing a literature review not mentioned above is that it provides a framework for relating new findings to previous findings in the discussion section of a dissertation. Without establishing the state of the previous research, it is impossible to establish how the new research advances the previous research.

However, writing a faulty literature review is, perhaps, the major way in which a thesis can be doomed. If the literature review is flawed, the remainder of the dissertation may also be viewed as flawed, because the candidate does not understand the literature and serious questions are raised about his or her ability to perform meaningful research (Mullins and Kiley, 2002; Boote and Beile, 2005; Randolph, 2009).

2.5.7 FINAL WORK

Many students complain that thesis writing is time consuming and frustrating. Indeed, writing a thesis can be frustrating but it is also a very rewarding experience.

Writing a thesis presents the student with the challenge and the opportunity of pursuing an intriguing intellectual question. It also allows the student to work in close proximity with an advisor.

Whatever the reason for and against writing a thesis, there is always a magnificent feeling of satisfaction when the job is done.

2.5.8 THESIS REVIEW COMMITTEE

The Review Committee is (preferably) composed of experts within the field of endeavor addressed by the thesis. But this is not always the case!

If the reviewers of a submitted thesis on, say, petroleum technology is composed of scientists and/or engineers with no experience in that area of technology, the purpose of the peer review system may be defeated. If the thesis is rejected because of lack of knowledge on the part of the reviewers, the system is defeated. If the thesis is accepted because of lack of knowledge on the part of the reviewers, the system is defeated.

2.5.9 THESIS DEFENSE

And then there is the thesis defense! Although this can generally take the format of a technical presentation (Chapter 9), the candidate must recognize that the form of the defense will be based on the thesis and how well the thesis is written.

During a thesis defense, the committee is going to determine (1) whether the candidate has a very good understanding of the problem, (2) whether he or she is conversant in the relevant literature and can explain how the proposed approach is different and more practical than existing work, (3) whether any assumptions are realistic, (4) whether the thesis work was a mere academic exercise or if it can be used in the real (nonacademic) world, (5) whether the candidate has taken the time to digest the results generated or whether he or she was merely a pair of hands for the supervisor/mentor, and (6) whether the candidate knows enough to have suggested future work and directions.

REFERENCES

Alley, M. 1996. *The Craft of Scientific Writing.* 3rd edition. Springer-Verlag, New York.
Boote, D.N., and Beile, P. (2005). Scholars before Researchers: On the Centrality of the Dissertation Literature Review in Research Preparation. *Educational Researcher,* 34(6), 3–15.
Day, R.A. 1994. *How to Write and Publish a Scientific Paper.* 4th edition. Oryx Press, Phoenix, Arizona.
Medawar, P.B. 1979. *Advice to a Young Scientist.* Harper & Row, New York.
Mullins, G. and Kiley, M. 2002. It's a PhD, Not a Nobel Prize: How Experienced Examiners Assess Research Theses. *Studies in Higher Education,* 27(4): 369–386.
Randolph, J.J. 2009. A Guide to Writing the Dissertation Literature Review. *Practical Assessment, Research & Evaluation,* 14(13): 1–13.
Sandia. 1990. *Format Guidelines for Sandia Reports,* Report No. SAND90-9001, Sandia National Laboratories, Livermore, California.
Speight, J.G. 2009. *The Scientist or Engineer as an Expert Witness.* CRC Press, Taylor & Francis Group, Boca Raton, Florida.
Strunk, W. 1918. *Elements of Style.* Humphrey Publishing, Geneva, New York.

3 Appearance and Physical Layout

3.1 ANATOMY OF A MANUSCRIPT

A well-written research paper involves giving a detailed description of a specific subject and, in addition to answering a specific question, statements in the research paper must be valid. Generally, the paper is written in three parts: (1) the Introduction section, in which background information is presented and related to the significance of the work; (2) presentation of supporting information, which includes the Experimental section and the Results section; and (3) the Discussion section, which includes supporting and opposing opinions, as well as the Conclusions section. Any other inclusions (such as an Acknowledgments section) are at the preference of the author(s).

As simple as the statements in the above paragraph may seem, writing a scientific or engineering manuscript is always a daunting task, and requires a great deal of planning, preparation, and time. To convey the experimental results in the clearest possible way, it is essential that a logical approach be taken in the formulation of each of the sections of the manuscript (Li, 1999).

Readers expect to find certain types of information in particular locations in a scientific paper (Table 3.1). Although the divisions between the sections are not cast in concrete, disregarding the typical layout can cause confusion among the readers—even those skilled in the scientific and engineering discipline.

The essentials of any scientific or engineering paper include a description of what is known, an assessment of what is unknown, a clear statement regarding the question and hypothesis being addressed by the current study, and a discussion and summary of new information that has been learned as a result of the study.

The approach described below is by no means the only way (and most certainly not always the very best way) to write a research paper. It is hoped that the following pages will provide scientific and engineering researchers with a formula by which they can approach the writing task and lower some of the hurdles that can make publication quite difficult.

These essentials can be covered in sections that are used as the backbone of technical papers: (1) the *Title*, (2) the *Introduction* section, (3) the *Experimental* section or *Methods* section, (4) the *Results* section, (5) the *Discussion* section, (6) the *Conclusions section*, (7) the *References* section, and (8) the *Acknowledgments* section (Table 3.1).

Finally, even before writing commences, the author should check the guidelines for submittal of papers to the journal. Many authors ignore the *crucial* step of reading and following the journal's submission guidelines. This is a serious omission and

TABLE 3.1

Anatomy of a Scientific or Engineering Manuscript

Section of the Paper (Skeletal Structure of the Manuscript)	Defining and Resolving the Problem (Fleshed-out Skeleton)
Abstract	Work that was done?
	What are the conclusions?
Introduction	What is the nature of problem?
	What do we know—prior work?
	What do we not know—prior work?
Experimental/Methods	How was the problem solved?
Results	What are the findings (data)?
Discussion	What do we now know?
	What is the meaning of the findings?
	What questions remain?
	Where do we go from here?
Acknowledgments*	Who gave assistance?
References?	What was the prior work?
Appendices*	Any other information?

* Not always necessary—optional extra.

can only cause frustration later when there are repeated requests from the Editor to the author to format the manuscript correctly. Failure to adhere to any such request from the Editor can lead to rejection of the manuscript, even after it has been judged technically sound and acceptable for publication by the reviewers.

3.1.1 TITLE

The title is the single most important phrase in the entire paper and has considerable impact. A potential reader who cannot extract the significance of a paper from its title is unlikely to read further. Shorter titles may be preferred by some journals, but they carry the risk of being too cryptic; although longer titles can be more informative, they are less likely to catch the attention of readers who scan quickly through journal contents or article listings.

The title should be informative, and any wordplay should only be used as embellishment. Again, when formulating the title, it is helpful for the writer to place himself/herself in the position of the reader and consider whether it sounds sufficiently informative.

Article titles are the means by which potential readers can recognize the content of the document. The title should use words and names that readers are most likely to look for in order to find the article in their search of the literature. The title should unambiguously identify the subject matter of the document. In addition, titles are expected to be shorter rather than longer—a full sentence title consisting of more than eight words is likely to deter many potential readers. If the title is overly verbose, they may argue that the document may be packed with unnecessary words.

Concise titles are preferred and titles using puns or clever wordplay, although not necessarily informative, can attract readers' interest, but this should not be done at the expense of information that portrays the focus and content of the article.

3.1.2 ABSTRACT

No matter what the nature of the scientific or engineering publication, an Abstract is, in the vast majority of cases, essential.

A scientific or engineering journal might state that the Abstract be no greater than 250 words. A company report may, on the other hand, require a multipage Executive Abstract or Executive Summary. Whatever the form of the Abstract or Summary, it must be tailored to the audience.

It is safe to assume that no matter what the backgrounds of the members of the projected audience, a meaningful abstract is essential. It is frequently on the merits of the Abstract alone that a reader decides whether to peruse a paper. The whole article might be a treasure trove of information, but if this does not come across in the abstract (or the title), the article has a good chance of being ignored.

A well-written Abstract should enable the reader to identify the basic content of a document quickly and to determine its relevance to the interests of the reader. The abstract should clearly state the objectives, scope of the investigation, and conclusions, which are not always obvious from the title. The Abstract should not include details of the experimental methods employed unless the study is method-oriented or is related to a new analytical method. In addition, it must be brief—generally less than 250 words or as otherwise defined by the journal guidelines. Many authors make the mistake of using too many words—if the essential details of the paper can be described in 100 words or less, do not use 200 and the information contained in the title should not be repeated in the Abstract. Furthermore, the Abstract should not contain any references to the literature and to tables or figures and obscure abbreviations and acronyms should not be used even though they may be defined in the main body of the paper. The Abstract, together with the title, should be self-contained as it is often published separately from the paper by the various abstracting services.

An Abstract differs from an Executive Summary (Chapter 2) because an Abstract is written for a scientific or engineering purpose. For a well-written Abstract, the author should write what is necessary, usually in the following order:

1. The author should prepare a pertinent sentence that encompasses the entire article or whatever he or she is summarizing.
2. The author should prepare two or three subordinate sentences that support the main idea or pertinent sentence.
3. The three or four sentences should be tied together to present the contents of the article using well-organized words and transition between the sentences.

For the scientific and engineering purist, there are two main ways to write an abstract: (1) free-form abstract and (2) a structured abstract.

A *free-form abstract* is usually written as a single paragraph, whereas a *structured abstract* is organized into sections, the most basic of which are objectives, methods, results, and conclusions (Hartley and Sydes, 1997). A structured abstract may, in some organizations, fall under the umbrella of Executive Summary.

3.1.3 INTRODUCTION SECTION

The Introduction section of the manuscript has three essential purposes, which can be accomplished by addressing the following issues: (1) that which is already known about the topic or *what we know*, (2) that which was is unknown at the commencement of the work described in the current paper or *what we do not know*, and (3) that which is new information and results from the work presented in the current paper or *what we now know*. This will give the reader a clear understanding of the nature of the current study, and he or she will be clearly aware of the context in which the study is being performed.

The *Introduction* should be very explicit as to the specific research in the document and contain citations to the appropriate already-published papers that are the background to the work. The Introduction should begin by introducing the reader to the pertinent literature. A common mistake is introducing authors and their areas of study in general terms without mentioning their major findings.

Thus, the first issue involves addressing what is known about the topic. To accomplish this, the author(s) should provide a comprehensive review of the major findings in the current area of study. It is important that the authors be as complete, fair, and balanced as possible in the assessment of the current literature. The more completely written this *what we know* section (i.e., the first part of the Introduction section), the easier it will be for the author(s) to describe *what we do not know* (i.e., the second part of the Introduction section). This segment of the Introduction will have set the stage for the *raison d'être* for the work.

The second issue (i.e., *what do not know*) involves identifying the gaps in the current understanding of the scientific or engineering field and the reasons (do *not* forget the reasons!) why it is important that these gaps in knowledge be closed. It is very important to identify and reference those studies in the literature that have addressed these issues or even closely similar issues, thereby allowing full disclosure regarding the novelty of the current work.

Many authors conclude the Introduction section with a clear statement summarizing *what we know*, *what we do not know*, and *what the paper aims to accomplish*.

On the other hand, some authors complete the Introduction section with a short paragraph detailing the essential findings of the paper using a phrase such as *In the current study, it is shown that . . .* followed by a short description of the methodology and evidence in support of (or against) the hypothesis. This focuses the manuscript and facilitates the way in which the remainder of the paper is written.

It cannot be overstated or overemphasized that omitting an important paper and thus overstating the novelty of the current study can be detrimental to a scientific or engineering author.

It is essential that the extent of the current knowledge be made to the reader—this can be remedied by a careful literature review *and* full disclosure to the reader. Failure to do so can make the work appear to (1) be incomplete or (2) hide important facts (thereby puffing up the importance of the current work), or (3) border on charlatanism.

Finally, once the pertinent literature has been introduced to the reader and the need for the current work has been demonstrated, the author should state clearly the scope and objectives. The Introduction can finish with the statement of objectives or, as some readers might prefer (this style will be made obvious by consulting papers perviously published in the journal of choice), with a brief statement of the principal findings. This sets the stage for the reader to determine (with the help of the title and Abstract) where the paper will take him/her as he or she to follows the development of the experimental evidence.

3.1.4 EXPERIMENTAL SECTION

The Experimental section (also often called the Methods section) in a scientific or engineering paper should provide the reader with a detailed analysis of the methods used in the performance of the experiments.

This section should provide sufficient detail so that the reader(s) can repeat the experiments and collect data that, within the limits of laboratory difference, fall within the accepted limits of *accuracy* and *precision* (see, for example, Speight, 2002, and Speight, 2005).

It is also important in the Experimental section to provide details of any control experiments; this section should list all of the necessary information about, for example, solvent composition and sequence information used in chromatographic separations. In fact, it is the Experimental section that reviewers turn to when critically evaluating the experimental design, and that subsequent investigators turn to when attempting to repeat the experiments. Any flaws in the experimental design can and will negate the whole paper and lead to rejection.

The usual order of presentation of the Experimental section, and the methods described therein, is chronological. However, similar experimental methods may need to be described together and strict chronological order cannot always be followed. If the experimental methods are new (insofar as there is a change to an established method), the author should provide all of the detail required to repeat the methods. However, if a method has been previously published in a journal or as a standard test method (such as is published by the American Society for Testing and Materials—the ASTM)—the name of the method and a suitable literature reference may be all that needs to be given.

Attention to detail is extremely important and there should be no attempt to *hide* or *forget* data that do not fit the author's hypothesis. Indeed, failure to acknowledge such data and the omission of a critical step in either the set-up or conduct of an important experiment can lead to screams of cheating (Speight and Foote, 2011).

When appropriately written, the Experimental methods section can and will provide an extremely useful resource for the respective scientific and engineering communities and speak to (support) the credibility of the author(s).

3.1.5 RESULTS SECTION

The results section is truly the heart and soul of the manuscript. It contains all of the data to support (or refute) the hypothesis that was proposed in the Introduction section. Many authors find it most useful to actually start the paper with this section and to build the rest of the paper around it.

The Results section is the section in which the findings are presented, with important trends described. The results of the investigation comprise the new knowledge that the author is contributing to science or engineering and the findings must be simply and clearly stated. The section should be free of excessive verbiage.

The results section serves to weave a coherent story and must communicate the findings to the reader in a logical, transparent manner. Therefore, it is perfectly appropriate to describe the results in a manner that makes sense, as opposed to describing the experiments in the order in which they were performed. It may be helpful to use subheadings to introduce new paragraphs, and to devote a paragraph to individual or closely related figures.

It is permissible to use the past tense when writing the results section, given that at the time of the writing, the experiments were indeed performed in the past (as opposed to the discussion section, where the present tense may be more appropriate).

When using figures, care should be taken to avoid simply repeating the findings in both the figure and the body of the text. It is important to avoid conjecture or speculation in the results section, unless a combined results/discussion section is being written.

Statistical analysis of the data can be included but there must not be corruption of statistics to enhance the results or to prove the hypothesis. Inclusion of any such misinterpretations can lead to loss of credibility of the author(s) as well as to claims of cheating.

Figure and table legends should provide a detailed description of the corresponding figure or table, within the space allocated. Care should be taken to ensure that each symbol in the figure (typically arrowheads, arrows, asterisks) or table (typically asterisks or footnote indicators) is explained fully. If a legend contains a description of the experimental equipment used to perform an experiment, care should be taken to avoid duplicating this description in the body of the text.

3.1.6 DISCUSSION SECTION

Some authors prefer to include the Results section and the Discussion section under one heading. The journal to which the manuscript is being submitted may condone or refute such a combination or may even be ambivalent.

The Discussion section allows the scientist or engineer to further expand on the Results section and (1) communicate the significance of his or her findings, (2) indicate how the findings support (or refute) the experimental hypothesis, and (3) describe how these results advance the field of study.

Nevertheless, it is advantageous to begin the Discussion section with a paragraph summarizing the main results, culminating in a statement describing the overall significance of the work. Subsequent paragraphs can be devoted to expanding on themes that the authors feel are important for the reader to understand the significance of the work.

One of the major issues that occur when writing the Discussion section involves overstating the significance or novelty of results. It is perfectly appropriate to make inferences about the significance of a set of studies in regard to the experimental system upon which the data are based. However, extreme care should be taken not to generalize the findings into other unrelated experimental systems.

At this point, criticism or praise for previous work (by other authors) should be included using extreme caution and only when the authors of the current paper have conclusive proof can they state that previous work was incorrectly performed (vis-à-vis the experimental work) or beyond reproach. Many aspiring and old-time authors have found this out to their detriment.

3.1.7 REFERENCES SECTION

An extremely important aspect of scientific and engineering writing is to build on the ideas of other scientists and engineers. The scientist or engineer needs to show that he or she has understood the materials that he or she has read/studied and that he or she is able to use prior ideas, hypotheses, and findings in a constructive manner. In fact, this is an essential skill for every scientific and engineering student.

A disturbing aspect of any published (or submitted) manuscripts is the belief (which many students inherit from their supervisor/mentor/professor) that providing a long list of references increases the validity of the article—this could not be further from the truth! The number of references should be reasonable and *relevant* (neither too many nor too few). In fact, some journals specify (rightly or wrongly) the number of references to be included in a particular type of study.

Prior work, in the form of a References List, should be cited and formatted as prescribed by the journal or publication to which the manuscript is being submitted. References are to be written correctly with due care and the correctly abbreviated, accepted names of the journals are to be used.

Referencing (the citation of published works) is used to support the claims of the writer throughout a paper or report. The citations in the text indicate the source of the information the author is using so that the validity of any conclusions can be easily checked. Referencing greatly strengthens its arguments and gives more validity to its substance. Authors should get into the habit of referencing all the data that go into their papers. If possible, original sources should be used—a textbook is only repeating information from earlier experimental papers. Mistakes in interpretation can occur between original experiments and textbooks, which rarely discuss the limitations of the experiments.

Company workers should also remember the value of referencing and citations. It is not advisable in an internal company report to include only minimal references to company-based reports when there are several hundred related references in the available scientific and engineering literature. Managers may only read the executive summary and, on that basis, decide to provide further funding for the work. Suddenly, many millions of funding dollars later, management discovers that similar work has been done elsewhere and is available through the published literature.

The methods of inclusion of references in a text are extremely variable. Whatever the format, it is important to properly and appropriately cite references in scientific research papers in order to acknowledge the sources and give credit where credit is due. Science and engineering move forward only by building upon the work of others. There are, however, other reasons for citing references in scientific and engineering research papers.

Citations to appropriate sources show that the writer has done his or her homework and is aware of the background and context into which the work fits, and they help lend validity to the arguments. Reference citations also provide avenues for interested readers to follow up on aspects of the work—they help weave the web of science. It may also be pertinent or necessary to include citations for sources that add relevant information to the current work or that present alternate views.

The writer should acknowledge a source any time (and every time) he or she uses a fact or an idea obtained from that source. Clearly, the writer needs to cite sources for all direct quotations and must cite sources from which facts and/or ideas have been paraphrased or summarized—whether or not the writer has put the fact or idea into his or her own words.

Sources that need to be acknowledged are not limited to books and journal articles, but include Internet sites, written correspondence, e-mail correspondence, and even verbal conversations with other people (in person or by telephone and acknowledged as personal communication with . . . ., accompanied by the day, month, and year). In fact, the author should seriously consider acknowledging all different kinds of sources. Furthermore, if the author uses illustrations, such as figures or graphical material either directly or in modified form, that he or she did not create or design, the sources of such illustrations need to be acknowledged.

There is a variety of styles used by journals for referencing information—citations in the text may be referred to by number or by author name and year. In the reference section the citations are then arranged (using the former system) numerically or (using the latter system) alphabetically by surname of the first author.

Numbered references (the former system) can be troublesome, especially of the work has been completed and an essential reference appears and must be included—many of the numbers will have to be changed as well as the numbers in the list of references, which often leads to confusion and mis-numbering.

The name–year system [e.g. (Speight, 2011)] is often the most convenient, with the references listed alphabetically by name of the first author in the list of references.

The name-year system is less troublesome that the numbered reference system. The late appearance of an essential reference can be overcome by an easy insertion and there is much less confusion because there is no need for renumbering the references.

However, whatever the system preferred by the writer, adherence to the journal requirements or the company requirements for citation of the references in the text is essential.

In general, it is important to be as inclusive as possible when referring to previously published work. Work that is in press should be indicated as such, although the purist would not include this in the references as the work is unpublished and, therefore, unknown at the time of submittal of the manuscript. Certain journals have stylistic restrictions regarding a difficult-to-obtain citation as *unpublished results* or the even more hidden data and conclusions in a *manuscript in preparation*.

3.1.8 Acknowledgments

The acknowledgment section allows for the acknowledgment of individuals who rendered assistance to make the work possible, and who are not coauthors. These may include mentors, administrative assistants, and individuals that proofread the manuscript. It is important to identify the sources of all reagents that were obtained as a result of collaboration. The sources of funding should also be acknowledged where appropriate.

3.2 PAGE LAYOUT

A scientific or engineering paper is a manuscript in which original research results are presented and discussed (Day, 1979; Alley, 1996; Matthews and Matthews, 2008). The format of a scientific or engineering manuscript has evolved through centuries of developing tradition, editorial practice, scientific ethics, and the interplay with printing and publishing services.

Scientists and engineers, in the infancy of their careers, often have difficulty producing documents that are well designed and consistent in appearance and compatible in layout with other documents relevant to their careers. As the scientist or engineer matures, he or she will attempt to incorporate a consistent format and style in their respective documents; however, unless there is an experienced mentor or peer reviewer, this is often a drawn-out process involving differences of opinion. Who has not had two or three peer reviewers with diametrically opposed opinions—especially in university departments where no one is responsible but everyone is subject only to his or her own authority?

To move beyond such petty squabbling, the young scientist and engineer should be prepared to follow the guidelines of the journal where the paper will be submitted or the conference where the work will be presented verbally through the medium of Power Point slides.

However, an interesting phenomenon may be observed on the company side in which the company's well-formatted templates bear little relationship to the well-formatted templates that exist outside of the company. In fact, aside from several formatting features that distinguish the look and feel of company documents from those of competitor companies, everything else is usually the same.

Since the advent of personal computing more than four decades ago, page layout skills have expanded to electronic media as well as to the print media. The young and inexperienced scientist and engineer may no longer exist having designed page-formatting techniques for assignment reports given as parts of university courses. They should, if the professor has any foresight other than his or her own personal quirks, have guided the students to use the format of a journal that is well known in the field of expertise where they now find themselves. Insisting on his or her own quirky formatting guides/instructions that are of little value or importance is not the way for a professor to guide and teach students.

Most organizations and journal publishers will recommend use of a predesigned template. Briefly, a template involves repeated elements mostly visible to the end-user/audience. Using a template to lay out the formatting elements of a document usually involves less skill than that which is required to design the paper from scratch. Templates are used for minimal modification of background elements and frequent modification (or swapping) of foreground content.

Nevertheless, the scientist or engineer must follow the template rigidly or suffer from delays in publication (assuming the peer reviewers are in agreement with the technical aspects of the subject matter) or outright rejection of the *unformatted* document.

More complex writing projects may require separate templates: (1) a page layout template and (2) graphics/table templates that describe actual design and size of the graphic illustrations and the tables. In such cases, the writing templates may suggest/request that the writer use an alternative page layout technology to include graphics and tables.

Furthermore, many scientific and engineering journals now use the final page layout for use in a *what you see is what you get* (WYSIWYG) version for publication in the journal, thereby reducing the costs of copy editing and production editing.

The format is the *layout* and *typography* of a document. Typography includes the style and size of type for a document. Layout includes the type of paper, margins, line spacing, paragraphing, and pagination (Sandia, 1990). However, there is no universal format for scientific and engineering documents. Each company or journal has its own format that suits the needs and desires of that company or journal.

Typography includes the size and style of type for a document. Type sizes are measured in points. In general, twelve-point type is used for the text portion of most documents. Larger sizes may be used for headings and titles, and smaller sizes may be used for footnotes and illustration call-outs. As far as the styles of types are concerned, two main classifications exist: serif and sans serif. The use of either depends on the situation, but a serif font of twelve points (such as Times New Roman) is generally accepted for the text portion of formal documents such as reports and correspondence.

Another aspect of typography is the use of initial capitals in titles and headings. The general convention is that the first letter of each word is capitalized except for articles, conjunctions, and prepositions that have fewer than four letters.

Finally, preformatting a manuscript in the print style and the layout of the journal anticipates acceptance and boasts perfection (or shows a thorough disregard for the Instructions for Authors). This will, most of the time, irritate both the Editor and the reviewers and may encourage rejection.

3.3 FONT TYPE AND SIZE

Use of a specified font size and type is not merely a matter of making the document look *pretty*; it is a matter of legibility.

Within the family of fonts, there are two main categories: (1) serif and (2) san serif—the serif font has the small features at the end of strokes within letters (see left example of Figure 3.1). Both groups contain faces designed for setting large amounts of body text, and others intended primarily as decorative. The presence or absence of serif forms is only one of many factors to consider when choosing a font.

Serif, or *Roman*, fonts are probably the most-used class in printed materials, including most books, newspapers, and magazines. Sans-serif fonts are commonly but not

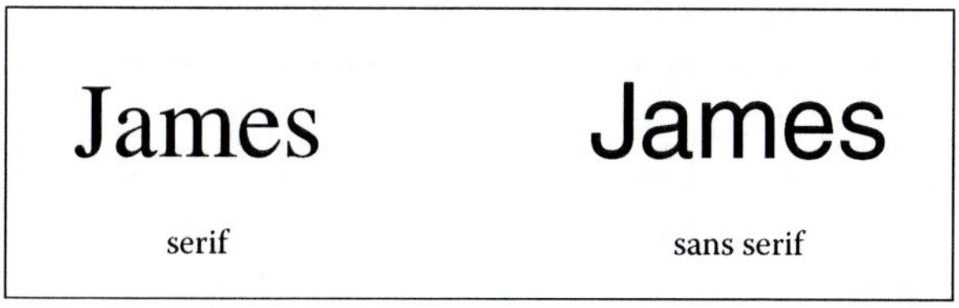

James James

serif sans serif

FIGURE 3.1 Serif (*left*) and sans serif.

exclusively used for display typography such as signage, headings, and other situations demanding legibility above high readability. The text on electronic media offers an exception to print: most Web pages and digitized media are laid out in sans serif typefaces because serifs often detract from readability at the low resolution of displays.

Fonts with serifs are often considered easier to read (by some readers, but it is a personal or organizational preference) in long passages than those without. However, most of this effect (or choice) may be due to the greater familiarity of serif fonts. As a general rule, printed works such as scientific or engineering journals and books use serif typefaces, at least for the text body.

For scientific and engineering documents, a twelve-point font is more legible than a ten-point font and serif fonts may support faster reading than the sans serif fonts. In addition, there is essentially no difference between the computer fonts and print fonts—in many modern journals, the correctly formatted and copy-edited manuscript is converted to a pdf format and then published in the next available issue of the journal. Many organizations are also using a similar procedure for publication of internal reports.

For the main text of the work, a font with serifs such as Times New Roman is often recommended. Most writers avoid the use of sans-serif fonts such as Arial or Helvetica—such fonts can make continuous reading difficult but this is a personal issue. The journal or organization will recommend the types of fonts to be used.

Although there is a wide range of fonts available, the author should choose one font and make it the consistent font for the work. Use of two or more fonts can be very confusing to the reader—the greater the number of different fonts used on a document, the less effective they become.

A given font size is not the size of a specific letter. Instead, it is the distance from top to bottom of the tallest ascender and lowest descender of the various characters in the font. Typically this includes a small space above the highest ascender and an equally small space below the lowest descender. The distance is measured in *points*—a point is typically defined as one seventy-second of an inch. In the designation of font size, the space between successive lines and the size of individual letters will vary.

In general, the size of the chosen font for a scientific or engineering manuscript should be twelve points. This makes the work easy to read, and the font will appear proportionate to its use when printed out on standard-size paper. The Instructions for Authors, made available by most journals, suggest the use of fourteen-point size for subheadings, and sixteen or eighteen point for main headings. Long quotations (where necessary) are normally set in italics and with larger margins on each side—if the quotation is given its own separate paragraph).

Finally, the writer should not use more than two fonts for the text and should also avoid nonproportional fonts (such as Geneva, Courier) as they are difficult to read in the form of long paragraphs.

Display fonts, a used for poster presentation and PowerPoint presentations (Chapter 9) are another issue entirely. They are chosen on the basis of readability but the reader or audience members are only looking at one or a few words and do not have to read (or scan) paragraphs of text. It is common to have sans-serif fonts for display text but experimentation and looking at professionally published posters or slides are useful aids.

3.4   TEXT ALIGNMENT

Text alignment specifies whether the text on the page is left-aligned, right-aligned, centered, or justified (left-aligned *and* right-aligned) and is used to describe the arrangement of text or graphics relative to a margin. *Left alignment* means that text is lined up along the left margin. *Right alignment* means that the text is lined up along the right margin. *Centered alignment* means that text is aligned around a mid-point of the page (or the midpoint between the margins). *Justified alignment* (commonly referred to as *fully justified*) means that text is lined up along both the left and right margins, that is, the text is left justified *and* right justified.

The Instructions for Authors or organization preferences for manuscript production may give guidance for alignment. Any available document that gives such guidance should be consulted at the first opportunity.

After the title and the main headings, which may be presented in a slightly larger font size than the body of the text, the most important element of any written presentation is the layout of the page. No matter what the content of the work, it will look better if is given plenty of blank space around the paragraphs. There should be no attempt to cram the maximum amount of text onto each page.

The *margins* should be generous—one and one-quarter inch at the top and each side (or more at each side) with a one and one-half inch margin at the bottom to accommodate the page number and any footnotes is usual for many journals and company reports. If the work is going to be presented in a folder or binder, there may be a request to allow an extra 0.25" *binding offset* (also called a *gutter*).

Many computers by default are set at *single line spacing*—the typical request from journals is for double line spacing (sometimes a journal may request one-and-a-half spacing).

The issue regarding full justification (in which both the left- and right-hand sides are aligned) is still undecided and is a journal or company preference. Full justification looks neat and tidy as an overall impression but it can cause larger areas of white space to appear in the text, caused by irregular spaces between the words. If left-side justification is chosen, this will leave the right-hand side ragged, but the spaces between the words will be regular. If in doubt, full justification usually offers more overall neatness on the page.

If the work contains items such as numbered lists, columns of figures, or anything else which is set off from the left hand margin, always use the TAB key or the INDENT command to position the item. Never use the spacebar—this will not give precise alignment. *Single indentation* is for regular paragraphs while *double indentation* is typically used for the longer quotations that would otherwise occupy more than two or three lines of the text in the work.

On the other hand, if the guidelines or Instructions for Authors suggest the use of double spaces between each paragraph, there is no need to indent the first line—this is only necessary when there are no spaces between the paragraphs. One good reason for having the double spaces, apart from its looking more attractive, is that such formatting assists the author and the reader to differentiate between paragraphs as a separate part of the argument or discussion.

Many journal editors and company document editors frown on the use of footnotes—if a comment is worth a footnote, then surely the comment is worth a paragraph within the text.

In laying out the pages, the author(s) should avoid creating paragraphs that start on the last line of a page or that finish on the first line of the next page (*widows and orphans*). The solution to this problem is to control the number of lines on a page so as to push the text forward. An extra-large gap at the bottom of a page looks better than an isolated single (or even double) line of text.

Italics are normally used to show emphasis—when something is very important. They are also used to indicate a word of foreign origin. They may also be used for book titles.

Once the document is in its final form, the print-preview facility should be used to help finalize the contents before printing. At this time, the writer can check the spacing of paragraphs and the appearance of the text as well as ensure that the titles, subtitles, and any section headings are set at the correct font size and weight.

Throughout the formatting and layout of the document, the golden rule is consistency throughout, providing the document follows the company guidelines or Instructions for Authors.

Finally, in reference to graphical objects (figures, charts, and tables), *alignment* (as expected) describes the relative position of the object to the margin. Most drawing programs allow the author to align two or more objects so that their tops, bottoms, sides, or middles are aligned.

3.5   SECTION HEADINGS AND NUMBERING

Headings are the titles and subtitles the reader sees within the actual text of much professional scientific, technical, and business writing. Headings are like the parts of an outline that have been pasted into the actual pages of a report or other document. Headings are used to mark off the boundaries of the major sections and subsections of a document and keep a document organized and the writer focused on the topic.

Headings are an important feature of professional scientific and engineering writing: they alert readers to upcoming topics and subtopics, help readers find their way around in long reports and skip what they are not interested in, and break up long stretches of straight text.

Headings are the titles and subtitles that are contained within the actual text of much professional scientific and engineering writing. Headings are often the parts of

TABLE 3.2

Choices in Format for Heading Hierarchy and Reference Listings

Format Issue	Purpose	Options
Heading Hierarchy	To rank information	Hierarchy by type size (18 points, 14 points, 12 points)
		Hierarchy by white space (3 spaces, 2 spaces, 1 space)
		Hierarchy by type style (**boldface**, ***boldface italics***, *italics*)
		Hierarchy by number (2.0, 2.0.1, 2.0.1.1)
Reference Listings	To give credit to sources	[Author, Year]: [Jones, 1998]
		[Numbered]: [1]
		[Abbr. author, abbr. year]: [JON, 98]

an outline that have been pasted into the actual pages of a report or other document (Table 3.2).

The use of headings aids in establishing the hierarchy of the sections of a paper to help orient the reader. Topics within a paper that have equal importance will have the same level of headings throughout the paper. Headings can also function as an outline to reveal the paper's organization.

Headings are an important feature of professional technical writing: they alert readers to upcoming topics and subtopics, help readers find their way around in long reports and skip what they are not interested in, and break up long stretches of straight text. Headings also help the reader to find the pertinent information quickly, which is very important when they may have a stack of papers to read in a limited time.

Organizations and journals use a specific style and format for headings in documents and reposts. For reports, many scientists and engineers must write according to an *in house* style and most organizations expect (dictate) that their documents look a certain way. Similarly, journals also expect submitted manuscripts to be formatted in the manner dictated in the Instructions for Authors.

The formatting styles used for section indentation and numbering give credence and importance to the various parts of a document. Managing numbered headings and outline numbering is anything but the simplest task and the writer can go round and round in circles, and never end up with what is required.

Whether or not the headings are numbered, a writer may want to use various formatting aspects of a document. For example, the second-level and third-level headings may have to be indented from the margin. For numbered headings, the second line of long headings have to line up with the first word of the heading, not the number.

While most word processing software has suggestions for styles, the journal Instructions for Authors, company guidelines, or book publishing guidelines will include suggested styles for the would-be writer. Nevertheless, the key is to use heading styles that are consistent with the company or publisher's directions.

Headings are also useful for writers. They keep the writer organized and focused on the topic. Some writers insert headings *after* he or she has written the rough draft. The correct procedure is to visualize the headings *before* commencement of the work and insert them, as the document evolves.

In the course of writing this book, a Table of Contents was designed and after acceptance by the publisher, each separate section of the Table of Contents became the outline for the relevant chapters.

Some general rules for the dos and don'ts of heading design and use are as follows:

- Make the phrasing of headings self-explanatory: instead of "Background" or "Technical Information" make it more specific, such as "Phase Separation during Hydrocracking."
- Make headings indicate the range of topic coverage in the section. For example, if the section covers the *design* and *operation* of a pressurized hydrocracker, the heading "Design and Operation of a New Hydrocracker" would be more informative.
- Avoid "lone" headings—any heading by itself within a section without another like it in that same section. For example, avoid having a second-level heading followed by only one third-level and then by another second-level—the third-level heading would be the lone heading.
- Avoid "stacked" headings—any two consecutive headings without intervening text are stacked headings.
- Avoid pronoun reference to headings. For example, if there is a third-level heading "Hydrocracking," do not begin the sentence following it with "This is a chemical means for producing . . ."
- When possible, omit articles from the beginning of headings. For example, "The Slurry Hydrocracking Reactor" can easily be changed to "Slurry Hydrocracking Reactor" or "Slurry Hydrocracking Reactors."
- Another omission is the use of words such as "Study of" and similar verbiage. For example, "Study of Hydrocracking Petroleum Residua" reads better when written as "Hydrocracking Petroleum Residua." The fact that a manuscript is submitted for publication to a journal or as an organizational report indicates that it relates to a "study of" the subject matter.
- Do not use headings as lead-ins to lists or as figure titles.
- Avoid "widowed" headings such as a heading at the bottom of a page and the text it introduces that starts at the top of the next page. Keep at least two lines of body text with the heading, or force it to start the new page.

As a final comment on manuscript anatomy, the manuscript for this book was prepared using Times New Roman 12-point font, with 1.25-inch margins at the top and at each side, and with a 1.5-inch margin at the bottom (so that the page numbers did not become associated with the text). The double spacing and the margins allowed sufficient space for clarity of reading and also space for comments by the copy editor. Section headings were discontinued after the second subheading and indicated using the point numbering system. For example:

1.0   MAIN HEADING
1.1   First Subheading
1.2   Second Subheading

The text of this manuscript was fully justified—this is a personal preference because full justification is considered (by the author of this book) to make the document easier to read.

However, the final word on heading and subheading and the means of identification of these heading will be found in the Instructions for Authors (if the manuscript is destined for a journal) or in the organization guidelines (if the manuscript is destined as an internal company report or report to a client).

Many journal guidelines or organizational guidelines may also specify full justification or only left-side justification of the text.

3.6   EDITING

Editing requires time and it requires a skeptical reading by the author/writer in which it is assumed that the document will be improved (Alley, 2000).

If a scientist or engineer reads his or her work while thinking, *This is well done and I will not find any problems*, the thought will be a self-fulfilling prophecy and the writer will not find many problems. On the other hand, if the work is read with the thought in mind, *I am certain that there are problems*, this is also a self-fulfilling prophecy: problems will be found, and the writing product will be improved. As a result, other scientists and engineers will gain a better understanding of the work.

Scientists and engineers are using computers for essay and project writing more than at any time in the past. The advantages for improved writing and presentation are dramatic. Once most scientists and engineers have started to take advantage of the facilities that computers offer for editing, rewriting, and presentation, they often wonder how they ever managed without them. As a result, typewriters (the mainstay of writing four decades ago) are regarded as an antique that has little use in the modern world of science and engineering.

The main advantage of the computer is that the scientist and engineer can rewrite and edit the product. A start might be made with a sketchy outline after which extra examples can be added, mistakes can be deleted, and paragraphs can be moved from one place to another. The finished product can be constructed in as many stages as are necessary.

The first draft is the key to the evolution of the paper. At first, many scientists and engineers opt to produce the first draft in handwritten form after which it is typed into a computer. The document can then be edited either on the monitor screen or by printing out the document and making handwritten notations for changes.

At first, many writers prefer to see what has been written in a printed form as soon as possible. As the young scientist and engineer gains experience however, he or she will probably edit on screen and only print out the finished version of the work. WYSIWYG (*what you see is what you get*) word processors allow the writer to see on-screen what the finished document will look like.

The final issue is proofreading. Editing and proofreading are two different activities. The former is an ongoing attempt to improve a manuscript by rewriting. The latter is a final check for mistakes of spelling, grammar, and fact.

REFERENCES

Alley, M. 1996. *The Craft of Scientific Writing*. 3rd edition. Springer-Verlag, New York.

Alley, M. 2000. *The Craft of Editing: A Guide for Managers, Scientists, and Engineers.* Springer-Verlag, New York.

Day, R.A. 1979. *How to Write and Publish a Scientific Paper*. 4th ed. Orynx Press. Phoenix, Arizona.

Hartley, J., and Sydes, M. 1997. Are structured abstracts easier to read than traditional ones? *J. Res. Reading*, 20: 122–136.

Li, V.O.K. 1999. Hints on Writing Technical Papers and Making Presentations. *IEEE Transactions on Education*, 42(2): 134–137.

Matthews, J.R., and Matthews, R.W. *2008. Successful Scientific Writing: A Step-by-Step Guide for the Biological and Medical Sciences.* 3rd edition. Cambridge University Press, Cambridge, United Kingdom.

Sandia. 1990. *Format Guidelines for Sandia Reports*. Report Number SAND90-9001. Sandia National Laboratories, Livermore, California.

Speight, J.G. 2002. *Handbook of Petroleum Product Analysis*. John Wiley & Sons, Hoboken, New Jersey.

Speight, J.G. 2005. *Handbook of Coal Analysis*. John Wiley & Sons, Hoboken, New Jersey.

Speight, J.G., and Foote, R. 2011. *Ethics in Science and Engineering*. Scrivener Publishing, Salem, Massachusetts.

4 Audience

4.1 INTRODUCTION

In composing any material intended for anything other than personal reading, an assessment of the audience is necessary. In the present context, manuscripts for submittal to a journal, an internal company report, a report to a funding agency, and slides for presentations all require an examination of the target audience (Gopen and Swan, 1990).

Knowing the audience for a written document or for a presentation will assist the scientist or engineer to determine what information should be included in a document or presentation, as well as how to convey the information most effectively. The scientific or engineering author should always consider the audience when choosing the tone, content, and language or the message may seem unfocused or inappropriate.

An audience is a group of readers (or listeners) who reads a particular piece of writing (or listens to a presentation). For most scientists and engineers, this catering to the audience is *the most important* consideration in planning, writing, and reviewing a document. The scientist or engineer must *adapt* his or her writing to meet the needs, interests, and background of the readers who will be reading the writing or to the listeners who will hear the presentation (Flower, 2000).

There are at least two types of audiences: real and intended. The real audience is anyone who reads or perceives the message (a message can be read or heard but not perceived—these two need not be the same); the intended audience is the target group that the writer has in mind. An audience addressed (the actual audience) versus an audience invoked (the perceived target audience) is basically the real audience versus the audience the scientist or engineer creates through the text and introduction. If it can be done, the writer tells the reader who he/she wants them (the members of the audience) to be.

Indeed, every scientist or engineer who decides to write a paper or make a presentation should anticipate the needs or expectations of the audience to convey information or argue for a particular claim (Berkenkotter, 1981; Flower and Hayes, 1981). The audience may be colleagues and peers, classmates, a manager, the president of an organization, or any other number of possibilities. The scientist or engineer needs to know the audience before writing commences. This requirement applies to communication of all kinds—from entertainment to education (Matthews et al., 2000).

The audience of a technical paper or, for that matter, any form of technical writing report is the intended or potential readership. For most scientists and engineers, the audience is an *extremely important* consideration in planning, writing, and reviewing a document. The scientist or engineer should adapt his or her writing to meet the needs, interests, and background of the intended readers. In fact, lack of audience analysis and adaptation is one of the root causes of most of the problems scientists and engineers find in professional documents.

All scientific and engineering writing hinges on knowing the audience—this applies not only to writing but to communication of all kinds (Chapter 2, Chapter 9, Chapter 10) (Matthews et al., 2000).

To understand the audience constraints, the writer must first decide whether the readers are scientists, engineers, managers, nontechnical persons, personnel from funding and donor agencies, or students. The writer must also assess what the audience knows about the subject in terms of what will need to be defined and the amount of background information that should be included.

Additionally, one must recognize that the target groups of scientists and engineers will want information with few frills and may be too narrow-minded to have a broader perspective. The members of this group will be obvious at a presentation as the question period is their forum.

Another consideration for the audience is the purpose of the document or presentation and to assess why the members of the audience are reading the document or attending the presentation. The writer's (or presenter's) final assessment will be to determine the reaction of the audience members to the document or presentation.

The purpose of scientific and engineering writing is to communicate findings (data) and ideas (theories or hypotheses) to an audience of readers (Flower and Hayes, 1981). Identification of the level of expertise and the interests of the target audience will assist the writer in identifying which aspects of the work should be presented and how the paper should be written. For example, an article on the operation of a refinery written for a daily newspaper or weekend magazine will have quite different content from a paper written on the same discovery for a journal such as *Petroleum Science and Technology*. First, the author should explain what a refinery does and why it does it while in the second the author should provide details of the composition of the crude oil to be refined and the type and character of the products.

When scientists and engineers talk to face-to-face, they automatically adjust their speech to be sure of communicating the message. However, many scientific and engineering writers do not make the same adjustments when they write to different audiences, usually because they do not take the time to think about the readers.

To be sure that scientists and engineers communicate clearly in writing, they need to adjust the message—the means by which they communicate the message and the information included—by recognizing that different readers can best understand different messages (Trimble, 2000).

Furthermore, when a paper is planned for a wide audience such as occurs with many papers submitted to learned journals, the writer should be prepared to identify the audience and have a better sense of the background of the audience members. This can be in the form of several questions (a checklist):

- What is the level of expertise of the audience in the general subject area?
- How familiar is the audience with the specific problem that will be addressed in the document or in the presentation?
- Which aspects of the results will be of greatest interest to the audience?

- Which aspects of the methodology or experimental setup will be familiar and of interest to the audience?
- Will the audience expect a particular format?
- Will the audience expect the text (presentation) to be written (spoken) in the first person or in the third person?
- If the product is a document, will the audience expect an essay, an article, a laboratory report?
- If the product is a document, will the audience need guidance through the document in the form of (1) a title page, (2) a table of contents, (3) headings and subheadings, (4) graphics and/or tables, and (5) a reference list?
- In a document or a presentation, how much jargon and which technical symbols will be familiar to the audience?
- In a document or presentation, will the audience require clarification or definitions of key terms?
- Should the document or presentation be addressed to a larger (or smaller) audience than had been originally considered?
- Will the audience expect outside sources to be cited? What types of sources? Sources from academic journals or popular magazines? Should Internet sources be cited?
- Will the audience expect personal experience to be included? In the introduction? As evidence in the body of the text?

Scientific and engineering audiences, like scientific and engineering writing (Alley, 1996; Matthews et al., 2000), tend to be highly structured, concise, and specific in their respective disciplines. Thus, specific formats are typically used to structure science and engineering writing, be it a lab report or journal article or a verbal presentation (using slides). In order to clearly convey information to the audience, sentences are typically short and pointed—wordiness can mask the point. Likewise, short words are always preferable to long words (the *Thesaurus* function is not appropriate for science writing).

It is a common mistake of novice scientists and engineers to waste too much time reviewing familiar facts while leaving too little time for the technical discussion of results. It is an equally common mistake to introduce too many details of the research too quickly. In either case, the audience will lose interest and fail to see the point of the work.

To familiarize himself/herself with the audience and style of a journal to which the writer plans to submit a paper, the scientist or engineer should first review the papers published in the journal of choice (or presentation made at similar conferences) and decide whether it is the correct medium for circulating the paper and if the readers or the attendees are the most suitable audience.

The presence of tables, charts, and graphs is a major difference between science and engineering writing and other types of writing. Tables, charts, and graphs are essential to good scientific and engineering writing and are an effective means of presenting results. In fact, it is only rarely that a scientific or engineering audience will not expect to find them in any report or paper or presentation.

The writer/presenter must appeal to the audience. At the first level, the appeal can begin with the choice of topic, according to what the audience might already know as well as what the *concerns/issues* of that audience are in that context. At the next level, audience appeal influences the type of proof that can be used to prove or persuade—part of an appeal to the audience is to use the type of information they would find most valuable. Appeals also involve how the scientist or engineer presents himself/herself—in other words, what kind of persona he or she creates on paper or in front of slides.

The scientist or engineer is ready to begin writing when the audience has been identified. This knowledge enables the scientist or engineer to select or reject details for that specific audience (Matthews et al., 2000).

However, no matter what the background of the individual members of the projected audience, a meaningful abstract is essential. It is frequently on the merits of the abstract alone that a reader decides whether to peruse a paper. The whole article might be a treasure trove of information, but if this does not come across in the abstract, the article has a good chance of being ignored. There are two main ways to write an abstract: (1) a free-form abstract and (2) a structured abstract.

A free-form abstract is usually written as a single paragraph, whereas structured abstracts are organized into sections, the most basic of which are objectives, methods, results, and conclusions (Hartley and Sydes, 1997). The latter type of Abstract may, in some organizations, fall under the banner of Executive Summary.

4.2   TYPES OF AUDIENCES

Reading scientific and engineering reports or papers or listening to a presentation is challenging. These works are written or presented in a very condensed style because of page limitations, time limitations, and the intended audience, which is assumed to be conversant with the relevant scientific or engineering area. Moreover, the reasons for writing the paper (or report) or making the presentation may make the audience work harder to find the content and to understand that content.

For this reason the scientific or engineering author or presenter must make sure that the work is understandable to the audience and such recognition is what requires that the author work to give the paper or presentation the correct approach.

Thus, one of the first things a scientist or engineer must do when he or she analyzes an audience is to identify the audience type (or types—it is rarely just one type). This allows the writer or speaker to adjust his or her work accordingly to fit the interests and capabilities of the audience. The common division of audiences into categories is as follows:

4.2.1   SCIENTISTS OR ENGINEERS

In this book, the term *scientists and engineers* refers to persons who wish to pursue or those who have already attained a master's degree or doctor of philosophy degree in science or engineering.

Science is regarded as inclusive of the life sciences, physical sciences, mathematics, and social sciences (the scientific study of human society and social relationships). *Engineering* is viewed as a field that includes all specialties such as civil, mechanical, electrical, petroleum, agricultural, and computer engineering. In all cases, the system of education of scientists or engineers should be organized around realistic experience.

In one way or another, a scientist is a person who has scientific training or who works in the sciences. On the other hand, an engineer is someone who is trained as an engineer. The practical difference between the two lies in the educational degree and the description of the task being performed by the scientist or engineer. It is widely believed that scientists explore the natural world and discover new knowledge while engineers apply that knowledge to solve practical problems, often with an eye toward optimizing cost, efficiency, or some other parameters.

The scientific and engineering fields are composed of educated and relatively young professionals who have the ability to apply themselves to the problems at hand, either theoretical studies or experimentation. To the scientist or engineer, the outcome of this work that offers some form of gratification is (1) completion of a project and (2) publication of the data in a journal or similar medium for distribution to one's peers. The latter gives the scientists or engineer professional recognition.

The various scientific and engineering disciplines are worldwide professional disciplines in which the members collect factual data and, through the ensuing treatment of the data, discover new arenas of knowledge that are universal. No one can foresee the difficulties that may emerge during a scientific or engineering investigation and know where experimentation and observation may lead. There is also the mode of data interpretation and such interpretation must be made at the highest ethical level with no attempts at embellishments, exaggeration, or covering up data that does not fit the author's hypothesis (Speight and Foote, 2011).

While research in the scientific and engineering disciplines offers the exhilaration of discovery, researchers have the opportunity to associate with (1) colleagues who have made important contributions to human knowledge; (2) peers who think deeply and care passionately about subjects of common interest, and (3) students who can be counted on to challenge assumptions. Scientists and engineers also have many opportunities to work with different people to explore new fields and broaden their expertise, especially where disciplines overlap.

Within the group of scientist and engineers there may be a subgroup generally referred to as *experts* (or even *expert witnesses*, remembering that such reports will be used in court before a nontechnical judge and jury).

The *experts* are the people who are well versed in the theory and the subject matter of the project. Often, the experts have advanced degrees and operate in academia or in research and development areas of the government and business worlds. The experts often serve on review committees appointed by finding agencies or within the higher level of technical management within a company. The communication challenge faced by the writer or speaker is communicating to an audience consisting of technician or students and experts.

Finally, even though scientists tend to explore the natural world and discover new knowledge while engineers apply that knowledge to solve practical problems, often with the objective of optimizing efficiency or some other parameters, it is well worth remembering that there may be considerable overlap between scientists and engineers. There are scientists who design and construct equipment and engineers who make important and lasting scientific discoveries.

4.2.2 Managers and Executives

Generally, managers analyze and solve problems at the interpersonal level and manage teams of employees while executives analyze and solve problems at the interorganizational and intraorganizational level and oversee organizational units and organizations.

Thus, for the purposes of this text, a manager is the person responsible for one functional area whereas an executive (such as a general manager) is responsible for all areas. Sometimes, most commonly, the term *general manager* refers to any executive who has overall responsibility for managing both the revenue and cost elements of a company or organization. This means that a general manager usually oversees most or all of the day-to-day operations of a company or organization (Sayles 1979a, 1979b).

The chief executive officer of a corporation retains overall accountability, while a chief operating officer may be delegated several responsibilities, including the authority to oversee other executives who direct the activities of various departments and implement the organization's guidelines on a day-to-day basis. In some companies or organizations, one or both of these persons may request to be the recipient of reports, documents, or attend presentations on a particular scientific or engineering project. It is doubtful that the president of a university may request to be the recipient of a scientific or engineering paper or to be present at a presentation by a faculty member.

The nature of the responsibilities of high-level managers and executives depends on the size of an organization. In small organizations, a manager or executive may be fully aware of the nature and intricacies of the various projects. In large organizations, where top executives are also responsible for implementing strategies and setting the overall direction of a certain area of the company or organization, being aware of the nature and intricacies of the various projects is less likely.

It is reasonable to assume that managers and executives know less about the technical details of a project than the scientists and engineers involved in the project. As a result, the scientific or engineering author of document or presentation will have to play the role of teacher and/or coach for the manager or executive. The extent of teaching and coaching (which should be determined beforehand by the scientist or engineer) depends on the background of the manger or executive. But a word of caution here—the scientist or engineer should not give the perception of talking down to the manager/executive or dumbing down the document or presentation. Background checks of both the manager and executive will often indicate the best approach.

Many scientific and engineering managers or executives who read a technical article or attend a technical presentation may have formal education and work experience as scientists or engineers. They should also have strong communication and business management skills for planning, coordinating, and directing research and production activities and may even supervise scientists, engineers, and technicians, alongside support personnel.

These managers and executives use their knowledge of engineering and natural sciences to oversee a variety of activities and determine scientific and technical goals within broad outlines provided by executives who are higher up the management ladder.

To perform effectively, these managers/executives must also apply knowledge of administrative procedures, such as budgeting, hiring, and supervision as they submit budgets for projects and programs and determine staff, training, and equipment needs. They hire and assign scientists, engineers, and support personnel to carry out specific parts of each project and supervise the work of these employees, check the technical accuracy of their work and the soundness of their methods, review their output, and establish administrative procedures, policies or standards—such as company polices and/or environmental standards. Such managers sometimes conduct their own research in addition to managing the work of others. Executives (even those with scientific or engineering training) do not usually have the time to participate in research programs but may be well aware of the basic principles of science and engineering.

However, some reports have to be written for a varied readership, for instance technical managers and financial managers. Writing two separate reports would be time-consuming and perhaps insulting as some managers might feel offended if they did not receive both versions! Strategic use of appendices and summaries (see Section 5.1) can produce a partial answer to this problem, but it must be identified as a possible area of difficulty before the writing starts.

The management title and background of the reader will determine not only the approach but also the technical level and the style of the writing. However, a balance must be maintained between the informality of friendly writing and the formality due to any business communication; diagrammatic material must be clearly produced even if the drawing is informal in style.

The objectives of a report are found in matching the reader's need, the information involved, and the appropriate tone, within the clearly defined limits of the report specification. When the objectives are identified and clarified in this way, the actual writing of the report will become much easier.

Since managers are busy people, the shortest report will generally attract attention and will also be read as opposed to being glanced at and forgotten. If the main text is uncluttered by detailed statistics, maps, explanations of technical terms, or experimental data, it will be kept as short, and therefore as readable, as the material allows.

An Appendix is often the most appropriate place for background information, which most readers will take for granted but of which a few will need to be apprised. Appendices are excellent for supporting statistics and diagrammatic material that are not needed while the report is being read. It is an ideal place to list symbols, technical terms, and abbreviations which may not be familiar to all readers. If few readers know them, they may be better placed near the beginning of the report.

As in many aspects of report presentation, tone (getting it right for the reader) is all-important. Managers may feel patronized if too much is explained and bewildered if too much is taken for granted. Putting helpful back-up material into an appendix can satisfy all parties. This is particularly important when the management readers have varied expertise.

The engineer may not want complex technical material to be interwoven with details of costing, but both technical and financial readers will be happy to find the latter in an appendix. The research scientist or engineer may be fascinated by the experimental details and will be pleased to find them in an appendix, while the report is primarily used by the marketing people who want results and not laboratory tests. Appendices are the report writer's friends but, while they are being developed, so is the main text of the report.

Summaries are different from abstracts and are used in different ways, although they are often confused.

The summary is the last part of the text to be written and it is perhaps the most difficult to write. The use of summaries is widespread, and they serve several purposes. A summary gives a general picture of the report for those who want to be reminded of what they have already read, and also for those who will never see—or want to see—the total report. Many companies have long circulation lists for summaries but only supply the full report to those whose status demands it or who ask for it (which most summary readers are happy not to do). In this way, summaries save much time and money and are therefore admirable.

Perhaps the most important use of the summary is by the most senior reader (an executive)—often the decision maker—who has neither time for nor interest in the detail; this reader wants the report's *answer*, its conclusions and/or recommendations, to be immediately available.

However, the writer has the unenviable task of trying to reproduce the balance of the original report in a very small space, while at the same time emphasizing the most important conclusions or recommendations that everyone needs to know.

4.2.3   Technicians

The technicians are the workers in the various fields of science and engineering who are proficient in the relevant skills and techniques and have a relatively practical understanding of the theoretical principles.

Experienced technicians in a specific technical area typically have intermediate understanding of the theory behind the work and expert proficiency in technique. As such, technicians are generally much better versed in technique compared to many faculty and general professionals in that field of technology.

Scientific and engineering technicians solve technical problems by (1) building or setting up equipment, (2) doing experiments, (3) collecting data, (4) calculating results, and (5) helping to make a model of new equipment. Some technicians work in quality control where they check products, do tests, and collect data.

In short, the technicians are the people who build, operate, maintain, and repair the equipment and design of experiments that the scientists and engineers theorize about. Theirs is a highly technical knowledge as well, but of a more practical nature. The abilities of technicians should *not* be underestimated because they are the people upon whose shoulders the success of many projects can wax or wane.

4.2.4   Nonspecialists and Nontechnical Personnel

Nonspecialist readers and listeners have the least technical knowledge of all. Their interest may be as practical as that of technicians, but in a different way. They want to use the new product to accomplish their tasks; they want to understand the new power technology enough to know whether to vote for or against it in the upcoming bond election. Or, they may just be curious about a specific technical matter and want to learn about it—but for no specific, practical reason.

Although many managers have scientific or engineering backgrounds, a particularly challenging assignment for the scientist or engineer occurs when he or she is confronted with a nontechnical management readership or audience. In such cases the systems of writing and presentation devised for technical audiences are defunct.

Scientific and engineering writing concerns the application of effective writing to lab reports, scientific articles, deviation reports, follow-up letters, site visit reports, and standard operating procedures for a management audience. Common problems that often accompany these documents include: a lack of clarity, verbosity, poor understanding of the function of the various documents and sections within documents, and lack of conciseness and flow.

Thus, a particularly challenging assignment for the scientist or engineer occurs when he/she is confronted with a nontechnical management readership or audience. In such cases, the systems of writing and presentation devised for technical audiences are defunct.

Explaining scientific advances to a nontechnical management audience (i.e., a nonscientific or nonengineering audience) is an interesting endeavor for the junior or mature scientist or engineer.

Assimilating a piece of research published in the primary scientific or engineering literature and reorganizing that information to produce a successful piece of science journalism is an excellent exercise in summarizing information, simplifying complex material, and de-jargonizing the writing. Most important, writing for a general audience of intelligent nonscientists and nonengineers is an excellent way in which the scientist or engineer can determine if he/she really understands a project concept.

The danger for the writer (or presenter) is that he/she slips into an *academic mode* or into *industry mode* and fails to focus on presenting the findings in a language that is mentally or verbally accessible to the average, nontechnical reader. The result is an end product that is not particularly enlightening.

The scientist or engineer cannot accept *on faith* that the audience will understand what is written or presented and questions cannot be asked by an audience that does not understand the nature of the work. The scientist or engineer must be able to write or converse in a manner that is understandable by the audience and must not talk down to the audience.

The scientist or engineer must digest, condense, and translate the information into plain English for the audience.

And for all his or her strength in the technical arena, he or she strives to remain somewhat naive so that the written or spoken word enables audiences to gain an understanding of the subject area and how it affects them or the world. In this way the scientist or engineer is able to sort through all of the information he or she has gathered and make it understandable. The scientist and engineer must digest, condense, and translate the information into plain English for the audience. Sometimes, he or she may rely on intuition and a prayer. He or she needs to organize the information into a navigable text or picture presentation so that the body of information is accessible to the uninitiated.

4.2.5   FUNDING/DONOR AGENCIES

Writing or presenting information to a funding or donor agency invariably involves submission of a written proposal and an on-site verbal presentation to the agency. This is usually preceded by a *request for proposal* (RFP) from the agency.

A proposal is a request for financial assistance for implementation of a project. For a scientific or engineering project, the proposal must not be a *shopping list* of the preferences of the writer. Each item on the list of requirements must be justified so that the funding/donor agency can decide if some or all of the items will be provided. The writer must know and be able to communicate the purpose of each request while showing the relevance of each item to the project. The proposal must be carefully formulated and designed.

Proposal writing is a skill that requires some knowledge and practice. The purpose of the proposal is to inform and to convince the agency that the work will follow the statement of work as presented in the RFP. The project proposal should be an honest document and not a *sales document* and should not be used to preach, boast, or deceive. If the writer/presenter is convinced that the concept outlined in the proposal is sound and should be supported, his or her project proposal should honestly report such reasons to the decision makers who will have a technical background and who will weigh its merits against other funding commitments/requests. The proposal should clearly indicate how and when the project will end or become self-supporting.

The project proposal must reflect any relevant background work already achieved by the scientists or engineers and should be logically presented. It must be demonstrated that the project is worthy of funding.

Above all, the proposal and any ensuing presentations must be tailored to the interests of the agency being approached.

A good proposal begins with a clear idea of the goals and objectives of the project and states why the proposed approach will provide a significant improvement over current practice.

The writer should envision what improvements the project will make, and then ask himself/herself what activities must be developed, what instruments will be needed, or what coalitions must be formed to make the desired improvements. Focusing first on the goals and objectives helps ensure that the activities are designed to reach those goals.

After the goals and associated activities are well defined, the writer needs to consider what resources (e.g., people, time, equipment, technical support) will be necessary as part of the request for funding. A better proposal is likely to result if the goals and activities are clear before resources are considered.

The proposal should also mention what work has been done in preparation for the project and describe specific attempts that have been made to try the proposed improvement on a small scale. Evidence of preliminary work demonstrates planning and commitment to the project and often indicates the project's potential for success.

Having listed the improvements that the project will make, the writer then identifies the activities that must be developed, the instruments that will be needed, or what coalitions must be formed to achieve the desired improvements. Focusing first on the goals and objectives helps to ensure that the activities are designed to reach those goals.

The proposal should also mention what work has been done in preparation for the project and describe specific attempts that have been made to try the proposed improvement on a small scale. Evidence of preliminary work demonstrates planning and commitment to the project and often indicates the project's potential for success.

When a proposal requests significant funds for equipment, it is helpful to consider alternatives and explain why the specific instruments chosen are more suitable for the project and why others, especially less expensive ones, are less appropriate.

Advice should be sought from people who have been successful in the proposal process. Such activities should be undertaken early in the writing process; seeking advice is not a last-minute activity.

If possible, have someone unconnected with the proposal read and comment on a draft—with sufficient time allowed for changes prior to the submission of the final version. This unofficial reviewer can help to identify omissions or inconsistent logic before the agency reviewers receive the proposal.

Finally, the project summary (abstract) is the first thing that Reviewers and staff of the funding organization will read. It should be written clearly and concisely in the space allotted and should outline the problem, its objectives, the expected outcomes, the project's activities, and the audience to be addressed. Since Project Directors use the Summary to choose Reviewers for the proposal, this is also the reviewers' introduction to the project. Considerable effort and thought should be spent in preparing a well-written summary.

4.2.6 STUDENTS

The student audience is like any other in that it has needs and expectations that need to be met. The scientist and engineer, as a writer and as a member of the technical community, is obliged to understand students and their respective subjects more deeply.

Many students are bewildered when they read a technical publication or listen to a technical presentation and, as for any audience, students need to understand what is written or being presented as the audience determines how it is written or presented.

The student audience is just like every other audience in one important way—to write an understandable text or to make an understandable presentation, the scientist or engineer as the writer/presenter must appeal to the group's interests and needs.

The student audience can represent different cultures with international students who may have a wide range of English proficiency. With the expansion of social networks and online recruiting, it is now more important than ever that scientists and engineers write clearly for international audiences.

The goal of any effective scientific and engineering paper or presentation for student audiences is to bring scientific or engineering data to the attention of the students. The longer the message, the more students will get lost in the details. In addition, there should be no marketing jargon, which is often not applicable and inauthentic. Simple language is key—complicated verbiage often causes students to tune out.

The author/presenter should decide what the students want to learn about the project and promote some of its unique aspects in the message that is taken away. The students should be encouraged to think about the project and the author/presenter can ask for feedback in the form of questions or e-mail correspondence. This will improve the message response rate and give students more information about the project.

Finally, it is important for the scientist or engineer to determine to which of the above six categories the potential readers of the document (or listeners to the presentation) belong.

4.3   AUDIENCE ANALYSIS

Audience analysis is a task all scientific and engineering authors and presenters need to perform early in a writing project. The audience is often referred to as the end user or client and all forms of communication should be targeted toward the defined audience.

Simply stated, audience analysis involves the processes of gathering and interpreting information about the recipients of oral, written, or visual communication. It involves the conceptions of the writer or speaker concerning the recipients of his or her communication, and it is vital to the success of the intended message that the writer or speaker be aware of the needs, interests, limitations, and expectations of his or her audience.

For most scientific and engineering authors, audience analysis is the most important step in planning a target document (or presentation). In order for the final product to be fully successful, the document (or presentation) piece must be aimed toward the intended audience and that includes (1) the knowledge level of the audience, (2) any prevailing opinions among the audience members, (3) the needs of the audience, specially the reasons why the audience members will read the document or attend the presentation, and (4) any demographic characteristics of the audience members.

Background-knowledge, experience, training is one of the most important concerns for the scientist and engineer. It is essential to know the extent of the knowledge, experience, or training that can be expected from the audience. If some members of the audience lack certain background essentials, it may be necessary to include such knowledge in the document or presentation. The question always is, how much?

Knowing the *needs and interests* of the audience will help in the planning of the document or presentation. The scientist or engineer needs to know what the audience is going to expect—including those members of the audience who may wish to use what they have read or heard. It is equally important to understand what the audience does not need to know—such as historical detail and trivia that might be uninteresting to one audience but vital to the interests of a different audience.

It is common in the world of funding reports and company reports for the scientist and engineer to find that the report is to be read by *more than one audience*. For example, the report may be seen by technical people (experts and technicians) and administrative people (executives). This creates a quandary but is can be resolved. The report can be written so that instead of all the sections being for all audiences (often, a bad choice), the report can be written so that each section is created strictly for the audience that would be interested in it. Headings and section introductions can be used to alert the audience about where to go to read matters of their own particular interest.

While there may be an audience that fits into only one category, there is the distinct possibility for *wide variability in the audience* background. This automatically creates a difficult situation—if the report is written for the lowest common denominator of reader, the report is likely to be cumbersome and tedious and will be a disappointment to the remainder of the readers. On the other hand, if the report is not written to that lowest level of background, that segment of the readership will be lost. Although it is not the only means of resolution of the issue, most writers appeal to the majority of readers and sacrifice that minority that needs more help. The most appropriate resolution might be to place the supplemental information in appendixes or insert cross-references to work that might be suitable for beginners.

Other *demographic characteristics* such as age groups, area of residence, gender and political preferences *might* have an influence on how the scientist or engineer should design and write the document. Audience analysis may also be complicated by at least two other factors: mixed audience types for one document, and unknown audiences.

Thus, an audience analysis is essentially a study of the needs and requirements of the audience members. The key is to gather as much information as possible so that the final product (paper or presentation) will be well received by the target audience. In terms of its importance, audience analysis ranks highly with scientific and engineering writers due to the nature of the research and the content of their writing. Therefore, it is particularly important for them to know who their audience is especially in terms of their needs and wants. What they know or do not know, their level of understanding, their perspectives and views, and their expectations are useful information to a writer or speaker.

Analysis of an audience is understanding the profile and characteristics of a group of people so that a paper or presentation can be customized for them. The analysis would help determine, among other things, the most appropriate style, tone, format, and amount of information required.

The analysis will enable the author/presenter to better focus on factors such as audience background, the situation, and context in which the final product will be delivered. Understanding the analysis will be a great help to the manner in which the document is written or presented.

4.4   AUDIENCE ADAPTATION

Once the audience has been analyzed, the writer then must focus on making sure that the work appeals to the audience and the members of the audience are able to adapt to the author's or presenter's words.

Writing or presenting *to* an audience involves using a set of standard procedures that give the author a superior chance of connecting with the readers, whether the readers are specialist or nonspecialist. These are as follows:

- Add information that the audience needs to understand and adapt to the document presentation. Key information, such as important background or a definition of key terms, must be included.

- Information that the audience does not need should be omitted. Unnecessary information can also confuse and frustrate the audience and make adaptation difficult.
- Change the level of the information that is at a too-high or too-low technical level—such as has been written for an expert audience rather than for a manager-executive audience, and vice versa.
- The author/presenter should use examples to help audience members to understand and adapt to the paper or presentation. Inclusion of examples (especially analogies) is one of the most pertinent ways to help an audience adapt, particularly when trying to explain a new technical concept.
- The use of graphics is key to help audiences understand and adapt to the nature of a paper or presentation, remembering that graphics for specialists and experts tend to be more detailed, more technical, and less attractive to the nonspecialist.

These suggestions are often used by professional technical writers to fine-tune their work and make it as readily understandable as possible. It is the accumulation of seemingly minor issues that add up to a document being difficult to read and understand. Even though the issues may seem to be trivial and, in some cases, petty, they all add up until after one final error the reader throws up his or her hands in frustration and decides that the document is not worth reading.

Finally, whether writing papers or presentations is an art or a science, there are some general principles to follow:

- The language should be clear, to the point, and the relevance obvious.
- The paper or presentation should move at a good pace and the author/presenter should not lose the attention of the audience.
- The audience should not have to wade through endless facts, figures, or *laundry* lists.

Writing for (or presenting to) any audience must allow the reader (listener) to understand and adapt to the material being addressed, as is the case with scientific and engineering manuscripts (Chapter 2 and Chapter 6). This means that the topic must be introduced carefully through a series of standard segments.

As a refresher from previous chapters (Chapter 2 and Chapter 3), commence with an *Introduction* that enables the reader to understand what they are going to be reading about. The audience needs to know why the information is important. After presentation of the *Methodology* and *Discussion* of the results, the writer/presenter should finish with a *Summary* or *Conclusions* section. The last two sections are necessary to refocus the attention of the audience on the important issues and the meaning behind them.

The author or presenter should always bear in mind that it is easy to forget the limits of the audience's understanding and adaptability to new ideas and concepts. Specific science and engineering subdisciplines may be foreign to many conventional scientists and engineers and the topic must be explained in language they will understand.

Occasionally writing the subject in one or two, maybe even three sections, can make more sense. While it is often thought advantageous to learn everything about a subject in a long write-up or presentation, sometimes a long write-up (or presentation) is unwieldy, the audience fails to adapt, and their attention span is limited.

Finally, when writing a scientific or engineering manuscript for any audience, the author should consider the beginning, the middle, and the end, and remember why the piece is being written and for whom. The writer must constantly ask himself/herself these questions and decide whether he or she is fulfilling these criteria.

REFERENCES

Alley, M. 1996. *The Craft of Scientific Writing*. 3rd edition. Springer-Verlag, New York.

Berkenkotter, C. 1981. Understanding a Writer's Awareness of Audience. *College Composition and Communication*, 32(4): 388–399

Flower, L.S., and Hayes, J.R. 1981. A Cognitive Process Theory of Writing. *College Composition and Communication*, 32(4): 365–387.

Flower, L.S. 2000. Writing for an Audience. In *Language Awareness: Readings for College Writers*. 8th edition. P. Eschholz, A. Rosa, and V. Clark (Editors). St. Martin's Press, New York. Pp. 139–141.

Gopen, G., and Swan, J. 1990. The Science of Scientific Writing. *American Scientist*, 78: 550–558.

Hartley, J., and Sydes, M. 1997. Are structured abstracts easier to read than traditional ones? *J. Res. Reading*, 20: 122–136.

Matthews, J.R., Bowen, J.M., and Matthews, R.W. 2000. *Successful Scientific Writing: A Step-by-Step Guide for Biomedical Scientists*. 2nd edition. Cambridge University Press, Cambridge, United Kingdom.

Sayles, L.R. 1979a. *Leadership, What Effective Managers Really Do and How They Do It: Effective Behavioral Skills*. McGraw-Hill, New York.

Sayles, L.R. 1979b. *Leadership*. McGraw-Hill, New York.

Speight, J.G., and Foote, R. 2011. *Ethics in Science and Engineering*. Scrivener Publishing, Salem, Massachusetts.

Trimble, J.R. 2000. *Writing with Style: Conversations on the Art of Writing*. 2nd edition. Prentice Hall, New York.

5 Preparing to Write

5.1 RECORDS AND NOTES

Keeping records and notes (usually in a notebook) is a necessity, and keeping the notebook safe and close by is an indispensable habit of laboratory or field work. It is the records and notes that form the basis of any scientific or engineering document.

In times past, laboratory and field records were kept in bound notebooks and each page was signed and dated at the end of the day. Many organizations now allow the use of unbound notebooks (in the form of a three-ring or multiring binder), electronic files, or other formats. Loose sheets should be dated and immediately added to the binder, in chronological order, to meet the standards of good science or engineering and to ensure integrity of the research and the data.

Electronic notebooks allow (1) easier input of scientific data, (2) uniform format for data recording, and (3) the ability for collaborators to share and add to the record. Commercial electronic notebook software varies in how much it simulates a paper notebook, but usually includes most (if not all and in some cases more) of the functions of a paper notebook. If personally identifiable and/or sensitive data are involved, appropriate privacy and security standards must be observed. In fact, the security of electronic records, including access to a particular electronic notebook, its contents, and authentication of entries in a notebook, is a fundamental issue that must be addressed by the organization.

For the work to be defensible (especially in the case of patent litigation), there must be a means of signing and dating such records so that prior ownership can be established. This is usually accomplished by the use of a bound laboratory notebook or a carefully protected electronic notebook.

In any case, the data should be recorded in ways that cannot be altered. Records also should be protected from destruction (ranging from fading caused by sunlight or flooding to computer crashes). Usually a backup or duplicate system is necessary.

Laboratory or fieldwork in science and engineering requires good record keeping, which promotes both accountability and integrity in research. Records of research activities should be kept in sufficient detail to allow another scientist or engineer skilled in the art to repeat the work and obtain the same results. Moreover, when it comes to the point where the laboratory or fieldwork is to be transposed to a technical document for publication or internal circulation, the records serve as a base from which to work.

The more that is recorded, the more there is to work with as a notebook is an indispensible source of memory (maybe even a source of power as it can include entries of overheard conversations, images and memory, notes from work, and quotations). The more vivid the detail, the more likely memories will be jogged when needed.

The purpose of a notebook is to document the impressions and passing thoughts or ideas and concepts that can be used in writing.

It is the place where ideas begin and is the perfect place to explore, and to experiment with language. It is the place where many scientists and engineers break the rules and forget about grammar, syntax, or sentence structure. However, legal issues can require that the notes be readable and understandable.

Maintaining a notebook is recommended for recording thoughts, emotions, concepts, and words related to the project. Records need not be included in the final product but they can provide useful and even unique perspectives.

Thus, when it is time to publish or present the results of the research, the data needed to support the conclusions and analyses will be at hand. After publication, the scientist or engineer may need to deposit the data in a data registry and share it with colleagues who wish to repeat his or her experiments or examine the work more closely.

Finally, good record keeping can help defend a scientist or engineer against false allegations of research misconduct (Speight and Foote, 2011). Misconduct allegations commonly arise when other scientists or engineers are unable to repeat published research and the underlying reason for this failure often arises because the original research was not described in sufficient detail. While good research records cannot prevent any scientist or engineer from ever facing allegations of misconduct, the records can help to refute such allegations.

5.2   ORGANIZATION AND FOCUS OF THE WORK

Writing is a form of communication, and communication is almost always directed at an audience. An awareness of the audience (Chapter 4) is critical in determining the approach that the document will take.

The purpose of a scientific or engineering paper is to communicate results and analysis to the scientific community or to the engineering community. The better a paper is written, the more readers it will attract and the more citations it is likely to receive. This alone should be sufficient to convince any scientist or engineer to put significant effort into his or her writing; unfortunately, this is rarely the case.

However, there are two kinds of scientific writing: that which is intended to be read, and that which is intended merely to be cited—the latter tends to be infected by an overblown and pompous style (Gregory, 1992). In spite of this warning, the bulk of scientific and engineering literature is still almost unreadable, and is usually only read by scientists or engineers with a vested interest in the subject.

Having a paper in print can mean the world to the young scientist or engineer or it can mean absolutely nothing. Other than the writer, it means that the journal reviewers found some indication of new knowledge in the writing. However, if readers do not read or consult the paper, the writer has accomplished nothing, other than an additional line in his or her résumé.

Thus, after the author has discovered the requirements for the anatomy of the manuscript (Chapter 3) from the receiving body (journal, funding agency, company management) and investigated the character of the audience (Chapter 4), it is time to write.

The art of writing a technical document (some would prefer to call it the *science of writing a technical document*) is extremely important to the scientist and engineer and the effective communication of the scientific and engineering information is as important as doing the experimental work. The most definitive form of this communication is the research paper that has been peer-reviewed and published in a journal. A well-written paper is a pleasure to read, and can be an inspiration to colleagues.

The wide acceptance of publishing results has helped improve the advancement of science and engineering by calling attention to discovery of new concepts and new processes. As a result, scientists and engineers must develop accurate or useful descriptions of the work to be included in any document.

Indeed, constructing a well-organized scientific or engineering document is the first step to improving accessibility and readability. A well-structured paper with no worthwhile results or a poorly structured paper with worthwhile results is unlikely to be published. Moreover, papers that are written in a poor style in terms of expression and grammar are unlikely to appeal to editors, reviewers, or fellow scientists, and are also unlikely to be published in a reputable journal.

However, before the title or first sentence is written, it is helpful to scrutinize the results carefully, as modern science deals less with demonstrating facts and more with interpreting and discussing results (Horton, 1995)—this is often referred to as prewriting (Rohman, 1965; Flower and Hayes, 1981).

A scientist or engineer might withhold the results as his or her work is fully justified and requires no further explanation (because it follows his or her rationale), but the reader might think otherwise. In fact, many queries raised by peer reviewers (Chapter 1 and Chapter 8) are buried in the author's assumption that what has become obvious to him/her *must* be obvious to the reader. This also means that experimental results should be excluded if they do not contribute significantly to the main message of the manuscript, regardless of how interesting they are. If discarded results are sufficiently substantial, they might form the basis of another paper. This then borders on ethical issues and loss of credibility by the author (Chapter 3 and Chapter 8).

In short, there must be a deliberate and intense focus by the author(s) so that the written or verbal message is not lost in the folklore written into the paper or inserted into the verbal presentation.

Focus is, and has to be, the main message in the paper. The effective scientific or engineering author writes about the topic at hand and does not stray from that topic. The reader is not confused and the writing holds the reader's attention.

In other words, if someone asked a scientist or engineer specifically what a paper was about, could he or she tell him/her what specific points were addressed? Although *focus* is about the content or theme, it also includes the details that enrich and develop that message. For scientific and engineering writing, these details are often identified as being facts or data.

Indeed, a paper or presentation should address one main question, and the failure to do this is one of the most common reasons not only for reviewers to reject a manuscript (Lambert et al, 2003) but also for readers and the listening audience to do so (Chapter 4). A good research question should be specific, novel, and of interest to the scientific community or to the engineering community, and will dictate the choice of journal and its readership (Perneger and Hudelson, 2004). Readers of highly specialized journals will be easily turned off by lengthy explanations of what is common knowledge in the field; conversely, readers of general journals might need more background information to be able to follow the arguments. When writing a paper, scientists should put themselves in the readers' position and assess the depth of detail from their point of view and in terms of organization and focus of the document.

Organization is the flow of ideas that develop the overall focus of a technical document. The effective writer establishes a plan that has a beginning, middle, and end. This involves commencing with an engaging introduction and leaves the reader with a satisfying conclusion. Thoughtful transitions are used throughout and the purpose of the document and members of the audience to which the document is directed are clear.

Scientific and engineering writing is persuasive writing based on logical conclusions drawn from evaluating information or data. As a result, it is necessary to present information in a logical progression, pausing along the way to point out the important points that convince the reader that the author's conclusions are accurate.

The *focus* of a technical document is strongly supported when the writer explicitly connects data and information to the overall theme. This data and information can take the form of quantitative data, references to tables or graphs, or previously known facts (with proper citations). These details are used to support the writer's focus. This will require that the writer work on developing a precise language and clear explanations, because distracting or vague statements make the *focus* hard to identify.

A well-organized paper begins with a clear introduction placing the work in a general context (or explaining why the scientist or engineer is writing about a particular topic) and the supporting information is unveiled logically and at the right times so that the reader doesn't lose interest. Transitions between informational points are based on their *conceptual* relationships. The piece closes with a sense of resolution, usually by summarizing the important points.

Organization is weak when information is arbitrarily presented without having a purpose or supporting the central idea (i.e., *focus*). The author should not arbitrarily include a graph at the end of the paper or load data into a table without discussing its importance or relevance in the text. If the organization is strong and the transitions are smooth, the reader might know what the author is going to say before it is said!

There are several ways to organize writing (comparison-contrast, deductive logic, point-by-point analysis, or chronological history), but the most conventional and easiest strategy in scientific writing is to start a paragraph with a single conceptual point and then support or illustrate that point using available information or data. It is possible to save the main point until the end of the paragraph, but this requires a great deal of skill to ensure smooth transitions between paragraphs and very complete development of the logic supporting the paragraph's main theme.

If a writer feels overwhelmed by the task of organizing, he or she should compose an outline that shows the logical progression of information or create figures and tables first and use them to organize the points that will develop the focus of the paper.

The organization of any scientific and engineering work should commence, even before the title or first sentence is written, with a careful scrutiny of the results since modern science and engineering deals less with demonstrating facts and more with interpreting and discussing results (Horton, 1995). While a scientist or engineer might hold his or her results in high esteem (because they were obtained by a logical series of experiments), a reader may not agree. In fact, many queries raised by peer reviewers are shrouded in the writer's assumption that what has become obvious to him/her through logical progression (Mr. Spock notwithstanding) will be obvious to the reader.

On the other hand, excessive experimental details in the results section or unwarranted reiteration of results in the discussion will leave the reader wondering about the main message.

Results should be presented in a coherent and organized way that follows a logical path rather than a chronological path. Research is rarely a linear process from observation to hypothesis to experimental proof and this is what is relevant for the reader. It is therefore paramount to study the results and organize them in a logical fashion before writing the paper; otherwise this disorganization will manifest itself in the paper and be noted by reviewers and also, if the paper is published, by readers. Furthermore, including extraneous results that provide little support for the main theme will dilute the message and confuse the reader.

In addition, a result must be presented before it can be discussed, and any results that do not add to the point being discussed should be excluded. It is possible to cross the internal divisions of a paper to enhance the message and to preserve the flow of arguments, but this should be done judiciously.

As an aside, this can mean to the scientist or engineer that experimental results that do not contribute significantly to the main message of the manuscript should be excluded! This is nothing short of misrepresentation and cheating (Speight and Foote, 2011) whether or not the discarded results might form the basis of another paper.

If other scientists or engineers read the paper but remain unconvinced to use or, at the very least, to verify the findings, the author still has accomplished nothing (other than an additional line in his or her résumé). The paper will not be cited and this can be a career-ending prospect in a world where *publish or perish* is the order of the day in many institutions.

Thus, the scientist or engineer must have two clear objectives: (1) there is a need to convince both the reviewer of the worth of the scientific or engineering contribution and (2) after publication, scientists or engineers must be attracted to the document. The reviewer(s) must be convinced of the value of the work to approve it for publication in the journal and the reader must be willing to cite the work in a favorable light leading to fame (or sometime notoriety) of the writer.

Thus, in preparing to write a technical document for publication, it is important to select the journal carefully through consideration of the target audience (Chapter 4) and the journal's own statement of its scope. Since the data that will be presented are well known to the author/writer, but not to others, there must be serious attention paid to the structure of the paper for clear presentation of the findings. The writing phase can consume several drafts of both text and figures. It is essential to avoid mistakes because referees lose confidence in the data and conclusions if numerous errors are observed.

However, even before the words of a draft are put to paper, there is often (to many scientists and engineers) the *prewriting stage* (Rohman, 1965; Flower and Hayes, 1981) in which the author's ideas are developed mentally or as a series of notes taken from laboratory notebooks or other sources. At this stage, some sense should emerge from the wealth of information that the author has accumulated, and this may not happen in distinctive stages

After the prewriting stage, every scientific and engineering writer, whether a beginner or experienced, feels some anxiety when the first sentence of a paper or the first slide of a presentation is being addressed (MacMillan, 1988; Goben and Swan, 1990). Accordingly, many writers compose the *Introduction* after finishing the first draft, as a more representative *introduction* can be produced with knowledge of what has been written in the draft. Other writers find this approach confusing as their philosophy is that the *Introduction* should be in the writer's mind whenever the research gets under way. No matter what the order of writing the Introduction—essentially a personal choice—even first drafts need an *Introduction.*

Writing a technical document (and planning a technical presentation) is a process both linear and recursive (Matthews et al., 2000). It is linear because effective scientific and engineering writers construct documents in well-defined and ordered stages, but it is also recursive, because at any point an author may need to return to a previous stage.

Compiling the material for a technical paper or technical presentation is a challenge for many, if not all scientists and engineers (Day, 1995; Alley, 1996). The very nature of their careers requires that material be assembled for technical documents and for technical presentations. While company reports and presentations require that the relevant content be included in a standard format, technical writing and technical slides will vary in the world outside of the company. Requirements will vary with the nature of the publication and with the essential qualities of those who are the recipients of the presentation (Day, 1998).

Writing logical and readable technical documents and designing technical presentations are the aim of many scientists and engineers. As the scientists and engineers mature, their respective skills in expressing ideas will be (should be) honed and each will develop his or her own style of writing an outstanding technical document or designing an excellent technical presentation.

Writing an outstanding technical document (and by inference, designing an excellent technical presentation) consists of three stages: (1) planning, (2) drafting, and (3) revising.

The writing work starts from a plan that indicates how the work will be approached. The document will have a theme and will also include the extent of the writer's knowledge of existing work related to the issue for discussion/the problem and his or her thoughts about/attitude toward the earlier work. The writer will also need to search for information that either supports (or refutes) his or her ideas. The result could be surprising if he/she is not up to date on current thinking that is already published in the relevant literature.

At this stage, the purpose of the document or presentation should be identified by clarifying the reasons for its creation and its specific objectives. Scientific and engineering documents are often written as answers to a specific problem. This is articulated in a problem statement. Since such documents are tools designed for use by their readers (or listeners), the writer/author should define the audience as well as the level of expertise of the audience members (Chapter 4). Assessment of the attitude of the audience toward the writer/presenter and the subject matter is also important.

When the purpose of the document and the audience have been identified (Chapter 4), the scientists or engineer should be able to determine the document type and any specific format elements.

Quality writing is well organized with the writer making a list of main points that he/she wishes to develop in his or her essay. These points should be directly related to the topic and should include basic statements that will support his or her opinion. At this stage, preparing an outline will help movement to the final product and, with the purpose and audience in mind, the graphics and tables of data can be outlined.

Having developed the main points, the writer can start working on the draft, remembering that the structure of a quality document or presentation contains three sections:

- The introduction (*this is what I know and will tell you*).
- The body of the document or presentation (*this is what I now know as a result of my work and this is what I am telling you*).
- The conclusions (*this is what I have deduced as a result of my work and this is what I have told you*).

Assuming that there are coauthors, drafting may be a collaborative effort. Using the outline and the preliminary graphics/tables, the writer starts work on a *first draft*—a rough working version in which ideas are put on paper. Grammar, style, and usage are not now of great concern, but it is essential that important reference information is available and that a format is followed that is appropriate for the document type and its purpose. Graphics/tables should be included to illustrate and condense the information in the document (a picture is worth a thousand words).

The *Introduction* prepares the audience for the topic discussion. A clear statement of what is known and current thoughts and beliefs is the cornerstone for writing an outstanding technical document or preparing a first-class technical presentation. The *body of the document* (or presentation) is the informative center of the work and presents the main points of supporting or proving the thesis. These points divide the body of the work into distinct paragraphs with evidence that support the author's ideas. The conclusion restates the thesis and summarizes the main points of the work.

Editing and revision is an important stage of any written work (for publication or presentation). Paragraphs and sentences should be edited to make them easier to read by improving clarity, conciseness, and coherence. The choice of words should be examined to ensure that they are appropriate for the purpose of the document while grammatical errors—including parts of sentences, types of sentence, usage, punctuation, and spelling—are corrected.

Revision does not merely relate to correcting grammatical errors or changing a few words—it involves critically seeing what has been written. If possible, the first draft should be *put away* for a day or more, even a week, before revision in three stages:

- The format is checked to ensure that it conforms to the (required) conventions for that particular document type and publication medium.
- Ensuring that the organization and focus of the material is correct.

- The author(s) should ensure that proper credit has been given to the sources and ensure that the citations follow the appropriate citation style; omitting reference to a source because the original paper appears in an obscure journal is not acceptable (Speight and Foote, 2011).

An additional stage requires that the finished product be read to ensure logical development of the information while avoiding spelling and grammatical mistakes (consistently using U.S. or U.K. English throughout).

Peer review is an integral step of document and presentation preparation and is usually the last major step. Writers of most technical and scientific documents ask peers to review manuscripts for accuracy, clarity, coherence, and appropriateness. A technical expert will review the document for technical content while an editor may review to ensure that it conforms to the organization's style and to correct any remaining problems (Strunk and White, 1979). A supervisor or a manager should also review the document to ensure that it achieves the organization's purpose and is appropriate to the audience. A company may have a requirement for legal reviews.

Pertinent points to be considered when a manuscript is to be submitted to a journal for possible publication are: (1) consider the audience of the journal and the scope of publications that appear in the journal before selecting any journal for the paper; (2) planning, drafting, and revising are required stages of the work, which has a theme and is divided into an *Introduction*, *Body* (experimental data and discussion of the data), and (3) *Conclusion* (finding and deductions). The final revision involves correction of grammatical and spelling errors; proper use of citations; and making the document/presentation aesthetically pleasing.

As scientific and engineering writers compose a document, they also monitor their current process and progress. In fact, at all stages of the work, there must be regular reviews of the document as the work progresses. Reviewing depends on two sub-processes: (1) evaluating the document and its contents, and (2) revising the contents where necessary.

Reviewing, itself, may be a conscious process in which writers choose to read what they have written either as a step to further writing or with an eye to systematically evaluating and/or revising the text (Flower and Hayes, 1981). These periods of planned reviewing frequently lead to new cycles of planning and rewriting.

However, the reviewing process can also occur as an unplanned action and may be initiated by an evaluation of either the text or through planning by the author(s) (that is, the revision of written as well as unwritten thoughts or statements). The sub-processes of revising and evaluating, along with generating new thoughts and ideas, allow changes to be made at any time during the process of writing.

5.3 TIMELINE/SCHEDULE

The timeline is the schedule for completing a writing project. This is essentially the manner by which a scientific or engineering author can strategize his or her writing project.

In fact, effective scientific and engineering authors construct documents in well-defined and ordered stages. However, at any point in the writing project, the author(s) may need to return to a previous stage without serious disruption to the flow of the document.

Writing tasks can be conveniently divided into several steps, which may vary with the habits and methods of the individual. The writer should be aware of the due date for the work as well as the dues dates for each step of his or her timeline. If the due dates change, it is the writer's responsibility to note any such changes and revisit the timeline accordingly.

The timeline also reminds the writer of the number of days he or she has assigned to each step.

It is important to schedule due dates for completion of each component of the work. The end point of the Timeline is the date of delivery of the product and is set by the Publisher or the Agency who is funding the work. There are four major steps requiring due dates, which should also include reminders of the number of days assigned to each step (already noted above):

Step 1: Planning and document design.
Step 2: Development of a preliminary outline to assist with the organization of
        the information, keeping in mind the purpose and audience.
Step 3: Revision and editing of the document.
Step 4: The peer-review process.

The first step is concerned with planning and document design and may or may not be collaborative. In this step, the purpose of the document should be identified by clarifying both the reasons for its creation and the specific objectives. Often, scientific and engineering documents are written as answers to a specific problem, which is articulated in a problem statement.

Scientific and engineering documents are tools designed to be used by their readers. Accordingly, the audience needs to be defined (Chapter 4) and at the same time, the level of expertise of the audience should be investigated as well as the reason why the audience should read the document. It is also important to assess the attitude of the audience toward the writer and the subject matter of the document.

Once the writer has defined the purpose of the document, the research problem, the audience, and document type, the relevant information should be assembled. The next stage is for the author to develop a preliminary outline to assist with the organization of the information. In addition, and with the purpose and audience in mind, the author should develop any graphic aids such as figures and tables, to display the data.

Like other stages, drafting a document may be a solo effort or a collaborative effort. In either case, a first draft should be written using the outline and preliminary graphics—mainly to get the ideas on paper in some semblance of order and logic. At this point in the process, the author(s) may not be overly concerned about grammar, style, or usage. However, important reference information should be included and it is preferable that the draft follow a format that is appropriate to the type and purpose of the document. The graphics can be used to illustrate and condense the information in the document—remembering the old adage that *a picture is worth a thousand words.*

The next stage involves revision of the document. The revision process is not merely correcting grammatical errors or changing a few words, it is critically rereading the work. It is often preferable to put the first draft away for a day or two—some authors prefer three weeks—and then revise the document in three stages: (1) check that the format conforms to the conventions used for the type of document, (2) revise for organization, and (3) revise the content, making doubly sure that proper credit has been given to the sources and, if pertinent, an appropriate citation style (specified by the journal) has been followed.

Editing is part of the revision process—paragraphs and sentences should be edited to make them easier to read by improving their clarity, conciseness, and coherence. The choice of words should be appropriate to the purpose and audience of the document. Problems and inconsistencies in grammar should be corrected, including parts of speech, punctuation, and spelling. Authors who are nonnative speakers of English may also need to perform a separate editing for specific types of problems in grammar and spelling, noting that U.S. English and U.K. English spell certain words differently (e.g., *sulfur* and *sulphur*).

The last major step for most technical documents is the review process. Writers of most scientific and engineering documents ask their peers to review the manuscript for technical accuracy, clarity, coherence, as well as relevance to the subject matter of the journal to which the document will be submitted for consideration for review, acceptance, and publication. In many cases, a scientific or engineering peer should review the document for technical content. Finally, a supervisor or a manager may review the document to ensure that it achieves the organization's purpose and is appropriate to the audience. There may also be the need for legal reviews to ensure that company secrets (such as proprietary technology and patentable work) are not leaked prematurely.

Such a system can help the scientist or engineer (of all ages) in many ways. It gives the writer a plan for completing the assignment, keeping up to date with the progress made (as well as any lack of progress), and keeping track of current status.

A realistic timeline is set for completion of each step within the boundaries stated for the completion of the work. There must be time allocated to revisions and corrections for each step.

Unfortunately, many scientists and engineers create unworkable timelines for writing projects or may be subject to daily interruptions. Last-minute timelines contribute to finishing the work too late and subjecting it to the distinct possibility of rejection.

5.4 AUTHORSHIP

Scientific and engineering publications (articles, abstracts, presentations at professional meetings, grant applications, and internal company reports) provide the main vehicle to disseminate findings, thoughts, and analysis to the scientific and engineering communities. These documents must be published in sufficient detail and accuracy to enable others to understand and elaborate the results.

For the authors of such work, successful publication improves opportunities for funding and promotion while enhancing scientific and engineering achievement and repute. The benefits of authorship are accompanied by a number of responsibilities for the proper planning, conducting, analysis, and reporting of research, and the content and conclusions of other technical work.

In the current context, an *author* is a scientist or engineer who has made substantial intellectual contributions to a study, and whose authorship continues to have important academic, social, and financial implications. In some cases individual authors do not appear on the byline, because the document is issued under the entire responsibility of the organization.

In the case of publications with multiple authors, one author should be designated as the *lead author*. He or she will assume overall responsibility for the manuscript and serve as the managerial and corresponding author, as well as providing a significant contribution to the research effort. However, the lead author is not necessarily the principal investigator or project leader.

Generally, the corresponding author is typically one of the several authors who worked on a paper or report that has been submitted to a journal or periodical for review and publication. When a group of researchers and authors work together on a paper, which is not uncommon in many scientific and engineering disciplines, they usually choose one person among themselves to submit the paper to a journal or other publication. The corresponding author is the person chosen within the group to be responsible for all contact and correspondence with the periodical in which they seek publication of the work.

Infrequently called a *coauthor designee*, a corresponding author is not necessarily more important than any of the other authors of a paper, but simply assumes other responsibilities when dealing with publishing that paper. If only a single author worked on a particular project, then he or she is by default the corresponding author for any findings he or she wishes to publish. This title can be important for projects or papers that are the result of a great deal of work from multiple authors, who all wish to have their work reviewed and published.

The use of a corresponding author prevents multiple authors from slowing down the review and publication process. Any questions or comments the journal Editor may have regarding the submitted work can be directed to one author, who may then discuss the correspondence with his or her co-other authors and provide a single, authoritative response. The corresponding author is typically responsible for reviewing any drafts and changes sent by the publisher. This is why the choice of author to act as the voice of a group is important, because any errors or incorrect changes that go to print are the responsibility of this individual.

A *coauthor* should have participated sufficiently in the work to take responsibility for appropriate portions of the content. Each coauthor is responsible for the content of all appropriate portions of the manuscript, including the integrity of any applicable research. An individual retains the right to refuse coauthorship of a manuscript if he or she does not satisfy the criteria for authorship.

Authorship credit should be based on: (1) substantial contributions to conception and design, or data acquisition, analysis, and interpretation, (2) document drafting or critically revising for important intellectual content, (3) final approval of the version of the manuscript to be published or circulated, and (4) the ability to explain and defend the study at scholarly or internal company meetings.

Guest authorship, which is granting authorship out of appreciation or respect for an individual, or in the belief that expert standing of the guest will increase the likelihood of publication, credibility, or status of the work, is generally inconsistent with the definition of authorship.

When a group has conducted the work, if the authorship is up to the group, the group should be clearly and formally defined as such, identifying each member; once established, the group name must be used unchanged. Otherwise all individuals having direct responsibility for the manuscript and fully meeting the criteria for authorship should be stated as authors, and the other members of the group should be listed in the acknowledgments.

The order of authorship on the byline should be a collective decision of the coauthors. Authors should be prepared to explain the order in which authors are listed. Some documents containing contributions of different authors (i.e., conference proceedings) may be edited by one or more individual persons that are responsible for the document as a whole (editors).

Authorship conventions may differ greatly among disciplines and among research groups. In some disciplines the group leader's name is always last, while in others it is always first. In some scientific and engineering fields, the names of the research supervisors may rarely appear on papers, while in others the head of a research group is an author on almost every paper associated with the group. Some research groups and journals simply list authors alphabetically. In some disciplines, the listing order is not considered important and alphabetical listing is the order of the day.

Many journals and professional societies have publishing policies (guidelines) that lay out the conventions for authorship in particular disciplines. These policies state that a person should be listed as the author of a paper only if that person made a direct and substantial intellectual contribution to the design of the research, the interpretation of the data, or the drafting of the paper. However, students will find that scientific fields and specific journals vary in their policies.

Simply providing the laboratory space for a project or furnishing a sample used in the research is not an adequate contribution for inclusion as an author, though such contributions may be recognized in a footnote or in a separate Acknowledgments section. The Acknowledgments section also can be used to thank others who contributed to the work reported by the paper.

On the authors' side, a frank and open discussion of how these guidelines apply within a particular research project—as early in the research process as possible—can reduce later difficulties. Sometimes decisions about authorship cannot be made at the beginning of a project and, in such instances, continuing discussion of the allocation of credit generally is preferable to making such decisions at the end of a project.

Decisions about authorship can be especially difficult in interdisciplinary collaborations or multigroup projects. Collaborators from different groups or scientific disciplines should be familiar with the conventions in all the fields involved in the collaboration. The best practice is for authorship criteria to be written down and shared among all collaborators.

Above all, it is unethical to omit the name of a coauthor who makes a significant contribution to the research (Chapter 4). It then becomes a question of the nature of the contribution by the proposed coauthor. Was the person a technician and operated a spectrometer and nothing else? Was he or she a high-level technician or a professional who operated the spectrometer and presented an interpretation to the other authors?

Answers to these and related question should prepare the way to clearly designate the authorship of the paper.

The citing conventions for authorship vary and should be discussed early in the research when there is more than one collaborator. The Acknowledgments section allows inclusion of assistance that does not qualify as authorship.

All contributors who do not meet the criteria for authorship should be listed in an Acknowledgments section. Examples of those who might be acknowledged include a person who provided purely technical help, writing assistance, or a department chair who provided only general support. Financial and material support should also be acknowledged. Groups of persons who have contributed materially to the paper but whose contributions do not justify authorship may be listed under a heading such as *participating investigators*, and their function or contribution should be described—for example, as *scientific advice, critical review of the manuscript*, or *data collection*.

Finally, a word on the order in which the authors are listed in a scientific or engineering paper or report (Fine and Kurdek, 1993).

When preparing a scientific article for publication, there is often the question of the order in which the authors should be listed. In some cases, the person who conducted the research (and in many cases, this is also the person who wrote the article) is listed first, while the supervisor is listed last. Many supervisors balk at this and dictate (rightly or wrongly) that he or she be listed first.

When there are more than two authors, however, the additional authors are usually added in the order of their contribution and/or seniority between the first and last authors. As this is the often-accepted order of authorship, it is obvious to the reader who did the work and who oversaw the research.

The long-accepted hierarchy of authorship places the senior author first and the most junior author last. Many modern authors prefer to list the authors alphabetically by last name. A journal may prefer one or the other while an organization may have specific guidelines for listing the authors.

5.5 REFERENCES AND CITATIONS

A critical aspect of the scientific or engineering process is the reporting of new results in scientific journals in order to disseminate that information to the larger community of scientists (Chapter 1).

Communication of the results contributes to the pool of knowledge within a specific discipline and often provides information that helps other scientists and engineers interpret their own experimental results. Most journals accept papers for publication only after peer review by a small group of scientists who work in the same field and who recommend the paper be published (usually with some revision). Along with this general aspect of the paper is the need (one might say the *obligation*) to cite the work of others.

Publications vary but, in general, constitute (1) a report and a record of the activity of researchers, (2) the references upon which the research is based, and (3) the manner of evaluating the scientific and engineering activities of individual researchers. The quality of publications, in terms of scientific integrity, is therefore essential for research to be conducted in an efficient and responsible way and for a transparent communication between researchers and society. This can only be accomplished by acknowledgment of the work of others that has already been published.

Citing information from other publications is common, even necessary. However, except in rare instances, these should not be verbatim quotes (i.e., copied out). In all cases, the concept or sentence used must be referenced to the individual who wrote/thought of it. The writer should not take the chance that he or she looks like he or she is trying to take credit for the original ideas of another scientist or engineer—this is plagiarism and cheating (Speight and Foote, 2011).

Review articles are very useful, but they should be read, digested, and synthesized by the writer rather than directly copied. However, a review article may appear as a chapter of a book *but* only with the permission of the original authors and the publisher, with suitable acknowledgment. If there are questions that relate to the identity of original author or the original publisher, or if the current author does not agree with the opinions expressed in the original work, it is better that the review not be reproduced for publication.

In the current context, the final section is *references* (*literature cited*) where the sources cited by the authors are listed in the format required by the journal.

It is essential to credit published papers for work mentioned in a manuscript. There are a variety of ways for citing references in the text with the style used depending upon the policy of the journal. In-text citations should refer to the reference list. Like citations, a variety of reference formats are used by different journals.

References to, and citations of, the research findings of others are an integral component of any research paper. The usual practice is for the writer to summarize the finding or other information in his or her own words and then cite the source. Any ideas or other information that are not the property of the writer must be substantiated by a reference that is cited in the text. Generally, in research papers, direct quotation and footnoting are not always practiced—it is more convenient to restate (often with rewriting) the author's ideas or findings and provide a citation.

Plagiarism is the use of the words, ideas, and images of others without citation and is not tolerated. This can be easily avoided by adequately referencing any and all information used from other sources.

In some circumstances, too closely paraphrasing the words of other scientists and engineers may be construed as plagiarism. In journal style papers, there is virtually no circumstance in which the findings of someone else cannot be expressed in the writer's own words with a proper citation of the source.

Science and engineering disciplines are flourishing because of communication that is a much broader concept than publishing. Hopefully, before the article is published, researchers will have had extensive discussions with their peers to share their views, ideas, and opinions in order to check the validity of their claims. Unfortunately, the advancement of knowledge has not appeared to be a top priority for many scientists and engineers in recent times. Instead, fame and fortune have become the focus of the researchers and data that have not been confirmed or data that have been made up have appeared in scientific and engineering journals.

While there are claims that gross scientific misconduct are assumed to be rare, subtler forms of unethical behavior are becoming more common (Speight and Foote, 2011). For example, misappropriated credit in publications can lead to some of the most contentious conflicts in the academic world. Currently, in academia, publication of research data has become more competitive because universities and organizations are more focused on intellectual property and rights of ownership. In addition, research that is sponsored by commercial entities is usually controlled and it is the commercial organization that determines whether and how results are published. It is no longer an academic issue.

More specifically, *conflict of interest* exists when an author (or the authors' institution), reviewer, or editor has financial or personal relationships that inappropriately influence (bias) his or her actions (such relationships are also known as dual commitments, competing interests, or competing loyalties). The most easily identifiable conflicts of interest are those that are the most likely to undermine the credibility of the journal, the authors, and of science itself.

In summary, credit must be given to work that is cited. The format of the citation is dependent on the policy of the journal. Plagiarism and misappropriated credit in publications are contentious issues, while sponsored research is now within the purview of the funders in that they determine when, how, or if the results are published.

5.6   BIOGRAPHIES AND RÉSUMÉS

Many journals require that a short biography (or résumé) of each author be submitted with the paper. In most cases, the biography (or résumé) is included at the end of the published paper and is less than 100 words.

A biography (or résumé) will include the basic details of the author and his or her education, present employment, experience, recognition for accomplishments, and professional society memberships.

Unfortunately, *lying on a résumé (misrepresenting credentials)* is another common, type of deception. For reasons unknown, researchers have been known to forge credentials, which can be either blatant or take more subtle forms. Some authors embellish their résumés a little while others embellish a lot. Others just *lie*. Most of the time, the lie is about their education and the author is in a position where he or she has to continue the lie(s) until caught.

The key to stopping such practices is to check résumés thoroughly—not just the address and telephone number—but by thoroughly checking every line item listed for education and employment. With access to the Internet, it is easy to check on an author's credentials. Without assiduously checking the facts and claims, it is impossible to determine who will include falsehoods on his or her résumés.

The indications (but not conclusive signs) of lying are that the candidate is (1) on the defensive when asked to verify his or her education, and (2) unable to produce evidence in support of his or her educational claims.

Fake diplomas are also used to misrepresent an applicant's educational attainment. A search of the Internet will produce several online *diploma mills* that are willing to provide the custom-replicated diplomas from any learning institution. The selling companies are part of a growing number of Internet sites where people can buy phony credentials from real universities.

One of the earliest cases discussed by the NSPE Board of Ethical Review (Case 79-5) was about an engineer who received a PhD from a nondescript (*diploma mill*) organization that required no attendance or study at its facilities. The engineer then listed the degree on all his professional correspondence and brochures and the NSPE board believed (or has ruled in the past) that when listing a PhD, there is no reason to identify the university from which the degree was obtained. The mere listing of the advanced degree is widely understood to convey an earned doctorate.

In fact, misrepresentation of credentials, particularly regarding education, is often practiced. Fake diplomas from real universities are also available for sale.

Whether the inclusion of a biography of each author in a scientific or engineering paper is of value remains to be seen. Some readers prefer such addenda to a paper and find them interesting while other readers are ambivalent or do not even bother to read the biographies.

The only words of caution that can be added, on the basis of the above paragraphs, is *reader beware* if such a person is being considered as a potential hire.

5.7   READER

The reader of any scientific or engineering paper has the right to organized, concise, and interesting writing. Scientific and engineering writers must understand how they are responsible for such collections of *faux pas* where their readers mentally trip on acronyms, lose themselves in a labyrinth of disjointed ideas, or get trapped in extra-long sentences from which they are able to extract themselves only after multiple readings.

Although mention has already been made of catering to the audience of any scientific and engineering work (Chapter 4), there are other aspects that require the writer to cater to the reader. These are worthy of mention here.

During every phase of the work, the writer should put himself or herself in the position of the reader. This should help him or her to determine whether the work is readable as well as reduce the various aspects of the work in which the reader's abilities are taken for granted (Gowers, 1974).

Furthermore, effective scientific and engineering writers use words that are specific rather than general (*fluid catalytic cracking* instead of *heating*). Effective writers vary the sentence length—for example, if one sentence has five words, then the next sentence should have more than five—and effective writers begin their sentences with different words.

Scientific or engineering writing is unlikely to be read for pleasure during leisure time. This reading will be related to duty and for information and will almost certainly be under pressure of work, and may even be when the reader is tired. The writer needs to create a better impression by making the reading as easy and agreeable as possible.

The writer should think first about the objectives of the work and its effect on the reader. Assume that the objectives of the work are

- To display knowledge and understanding of a topic; or to expose misconceptions so that they can be corrected.
- To demonstrate that a task has been completed effectively.
- To present a convincing case.
- To explain an unfamiliar idea and perhaps even initiate a controversy.

The writer must, after carrying out these objectives, check them against his or her perception of the reader's requirements and ensure that they match. In particular, the writer needs to assess the reader's technical background so that the terms and concepts are appropriately presented.

In much of the work, a scientist or engineer may think that the reader already has a clear idea of the concept in the paper or the discussion in the slides. This thought should be accommodated during the planning and writing but should not dominate the work to the extent that it encourages writing something that he/she either does not believe or does not understand. Having planned the objectives and assessed the readers' requirements, the ideas or concept should be presented clearly.

When the author has submitted the *perfect manuscript*, there is an anxious wait until the journal editor responds with a decision. Immediate acceptance may occur but is not always likely. Rejection is common, but the author may be given an opportunity to revise and resubmit the paper, which is often improved by the referees'/reviewers' suggestions. If the editor is satisfied with the response, the work will finally be published.

The next letter is much more fundamental in that it is likely to signal acceptance or rejection—this letter brings either elation or devastation. All papers are important to their authors and there is no middle ground between potential acceptance and outright rejection. With luck (and good writing), the author will be given an opportunity to undertake revisions and resubmit the paper, often improved by the referees'/reviewers' suggestions—and if the editor is satisfied with the response, the work will finally be published.

Thus, it is advisable that, throughout the work, the writer puts himself/herself in the position of the reader to ensure that the work is clearly satisfying the needs of the particular audience.

REFERENCES

Alley, M. 1996. *The Craft of Scientific Writing.* 3rd edition. Prentice Hall, Upper Saddle River, New Jersey.

Day, R. A. 1995. *Scientific English: A Guide for Scientists and Other Professionals.* 2nd edition. Oryx Press, Phoenix Arizona.

Day, R. A. 1994. *How to Write and Publish a Scientific Paper.* 4th edition. Oryx Press, Phoenix, Arizona.

Fine, M.A., and Kurdek, L.A. 1993. Reflections on Determining Authorship Credit and Authorship Order on Faculty-Student Collaborations. *American Psychologist*, 48(11): 1141–1147.

Flower, L.S., and Hayes, J.R. 1981. A Cognitive Process Theory of Writing. *College Composition and Communication*, 32(4): 365–387.

Goben, G., and Swan, J. 1990. The Science of Scientific Writing. *American Scientist*, 78: 550–558.

Gowers, E. 1974. *The Complete Plain Words*. Penguin Books, Harmondsworth, London, United Kingdom.

Gregory, M.W. 1992. The Infectiousness of Pompous Prose. *Nature*, 360: 11–12.

Horton, R. 1995. The Rhetoric of Research. *British Medical Journal*, 310: 985–987.

Lambert, V.A., Lambert, C.E., and Tsukahara, M. 2003. Basic Tips about Writing a Scholarly Manuscript. *Nursing & Health Sciences*, 5: 1–2.

Perneger, T.V., and Hudelson, P.M. 2004. Writing a Research Article: Advice to Beginners. *International Journal for Quality in Health Care*, 16: 191–192.

Matthews, J.R., Bowen, J.M., and Matthews, R.W. 2000. *Successful Scientific Writing: A Step-by-Step Guide for Biomedical Scientists*. 2nd edition. Cambridge University Press, Cambridge, United Kingdom.

McMillan, V. 1988. *Writing Papers in the Biological Sciences*. Bedford Books, New York.

Rohman, G. 1965. Pre-Writing: The Stage of Discovery in the Writing Process. *College Composition and Communication*, 16(May): 106–112.

Speight, J.G., and Foote, R. 2011. *Ethics in Science and Engineering*. Scrivener Publishing, Salem, Massachusetts.

Strunk, W., and White, E.B. 1979. *The Elements of Style*. 3rd edition. MacMillan Publishing, New York.

6 Writing Style

6.1 GETTING STARTED

A *writing style* is the manner in which a scientist or engineer writer chooses to write a technical document either for publication (in a journal or as a presentation) or for circulation within the organization. The style not only reveals the personality of the scientist or engineer but also shows how he or she has analyzed the audience and adapted the document accordingly. The writing style reveals the choices the writer makes to express his or her thoughts (Strunk, 1918; Strunk and White, 1959).

The main issue of writing style is that it is subjective, and while different readers have different ideas about what constitutes good writing style, these differences are negated to some extent by the dictates of the journal or the company policy that oversees internal technical reports. Above all, the scientific or engineering writer must use a style that allows the audience to understand the technical issues because they are expressed (by the writer) directly, elegantly, and persuasively.

Scientific and engineering authors usually avoid the figures of speech and use precise descriptions instead of colloquial terms that might be found more often in more familiar forms of writing, such as text messages or e-mail messages. In a technical document, the scientist or engineer must explain how the work and conclusions relate to other published works and must also attempt to persuade the reader of the validity of his or her conclusions.

A scientist or engineer may decide to combine this with a personal style that places him/her at the center of the writing but it is generally accepted that personal style should only be used infrequently or not at all in scientific and engineering writing. A scientific or engineering paper/document/report that relies on a first person narrator, for example, may cause the reader to question its seriousness or the objectivity of the work.

In science and engineering, different writing styles are used to format various types of manuscripts and reports. There are styles used (1) in academia, (2) in government, (3) in the private sector, and (4) for journals that accept manuscripts for publication. Furthermore, within each of these three categories, styles will vary depending upon the dictates of the organization.

Scientists and engineers who are unaware of various styles of writing usually fail to write a document paper successfully.

Writing style is often difficult to define by many scientists and engineers and the definition is lost in catchphrases such as "I cannot describe writing style but I will know it when I see it." Such advice from a professor/mentor/supervisor/manager/boss is absolutely useless. It will, more than likely, leave the young scientist or engineer lost in the cacophony of words and arm-waving and will cause many hours to be spent wrestling with the document and the reality of several rewrites, all because the person in charge cannot describe what he or she requires in a document.

Help is available in the form of various organizations such as the Modern Language Association (MLA) (Gibaldi, 2009) and the Chicago Manual of Style (CMS) (CMS, 2010).

The Modern Language Association style is one style for writing a scientific or engineering paper. The standardized rules of the MLA include specific rules for pagination, headers and footers, bibliography, and citations. The Chicago Manual of Style (CMS) is applied to writing many scientific and engineering documents, and this style also includes guidelines for standardized citations.

The Modern Language Association style specifies guidelines for formatting manuscripts and using the English language in writing. It also provides writers with a system for referencing their sources through parenthetical citation in their text and *works cited* lists. Scientists and engineers who use Modern Language Association style correctly can build their credibility by demonstrating accountability to their source material. Most important, the use of the Modern Language Association style can protect the scientist or engineer from accusations of plagiarism, which is the purposeful or accidental uncredited use of source material written/published by other writers.

The Chicago Manual of Style also specifies guidelines for formatting manuscripts and using the English language in writing. It also provides writers with a system for referencing their sources through parenthetical citation in their text and *works cited* lists using two basic documentation systems: (1) notes and bibliography and (2) author-date. Choosing between the two often depends on subject matter, the nature of sources cited, and the publication medium as each system is favored by different groups of scholars.

For the scientist or engineer who wishes to acquire more information on different paper writing styles, there is a variety of other sources. However, the importance of using a recognized writing style cannot be ignored—using an appropriate style makes the finished document more reliable. Therefore, the significance of using a recognized writing style is evident in every form of paper writing.

Generally, a junior scientist or junior engineer (even a mature scientist or engineer) is puzzled and may even be confused when he or she first starts to compile documents related to his or her work. The major issue is how to start followed by a series of issues, not the least of which is the issue that style is subjective and the audience may have different ideas about what constitutes good writing style. For example, use of the passive voice is generally more acceptable in the hard scientific and engineering disciplines than in the social sciences. Neither of these approaches is incorrect; it is a matter of what is acceptable to (and understandable by) the target audience.

In addition, scientific and engineering writers usually avoid *figures of speech* and *clichés*. Scientific and engineering writing require exact descriptions rather than colloquialisms that might be found in text messages, personal memoranda, and personal notes.

Figures of speech are phrases that generally and briefly describe a complicated concept through connotation. If scientists and engineers have to resort to use of such phrases, the point of their work will not be taken seriously and will be lost to the reader because of insincerity and, on occasion, lack of meaning. A *figure of speech* may also be referred to as a *cliché,* and whatever the name used to describe such terminology, the scientific or engineering reader who is a member of the target audience may, very likely, have a strong opinion that such terms are better omitted from a serious technical document.

A scientist or engineer should not write in the same manner that he or she speaks to friends or colleagues during the coffee break. This will only serve to distract the reader from what is being written by *how* it is being written. In fact, if the scientist or engineer decides to use a personal style that places him/her at the center of the narrative, he or she may have failed to acknowledge the expectations of the audience. It is preferable that elements of personal style should not be used in scientific and engineering writing situations.

For example, a scientific or engineering paper that relies on a first person narrator may/will cause the reader to question its seriousness or objectivity. It may even appear to the reader as being written in the form of a fictional account of a technical event, leading the reader to doubt the author's credibility, and failing to keep the reader interested.

Learning how to recognize matters of style in writing will give the scientist or engineer more control over his or her writing and the manner in which the target audience reads and accepts the paper will be a result of choices of style that have been made. If those choices are deliberate, the writer will have more control over how the reader reacts to the technical argument.

Thus, the scientist and engineer needs to tailor his or her writing style to the situation. Put simply, the writing style is the effective means by which a scientist or engineer transfers his or her thoughts (in writing) to the reader/audience in a readable and understandable format (Strunk and White, 1959). This includes (1) choosing a format that is adequate to the task, (2) ensuring that the writing is clear, concise, and understandable, (3) describing the point of the work and presenting the conclusions in a crisp and understandable manner, and (4) minimizing the potential for verbal overkill in the document.

Finally, the goal in scientific and engineering writing is not to *sound* intelligent but to get an intelligent point across. The scientist or engineer may have read complicated books—which use multisyllabic words (sometime referred to as *pompous words*) that add little to the writing but take up space—and may try to imitate this type of writing, but this can cause the writer to miss the most important goal, which is to communicate his or her hypothesis and be understood. If the audience cannot understand what has been written by the scientist or engineer, the work is ignored and lost.

Once a scientist or engineer learns to recognize style in his or her writing, he or she will have more control over the writing and the manner in which the document is read by the audience will be more appreciative.

The most important goal in every paper/report is to get the point across in a manner as straightforward as possible and make sure that the audience understands it.

Some of the more common, style-related writing issues are presented below and may or may not offer the scientist or engineer a choice. In any case, the finished document will be presentable to an audience and in future works the author should find that he or she is becoming accustomed recognizing and correcting these issues as he or she writes.

6.2   SELECTING THE FORMAT

Apart from the audience (Chapter 4), an extremely important aspect of scientific and technical writing is the format of the document, which is the arrangement of type upon the page and also serves as a contributor to *style*. The rules for formatting a paper involve specific line spacing, font size, font type, and way of giving referencing.

However, given the wide variety of formatting guidelines that arise from the multitude of different journals, meeting presentations, and company reports, no one set of format guidelines can possibly present every format option that a scientist or engineer will encounter. It is not important that the scientific or engineering writer know every format that exists (an impossible task), but he or she should realize that a specified format will often exist and that format must be followed. For those situations in which no specified format exists, the writer should choose a professional format to follow that is appropriate for the situation.

As the scientific and/or engineering writer collects his or her thoughts about the document, a new element enters the task environment, one that may even place more constraints upon the writer. Just as the discussions of the data determines and limits the choices of what can come next, the influence that the growing text exerts on the composing process can vary greatly. This is the point where writing style plays a major role in the appearance and acceptance of the final document.

The choice of typeface, the placement of headings, the method of citing references—these are aspects of format, which all contribute to style (Table 6.1). For longer documents such as reports, format also encompasses the arrangement of information into sections. In science and engineering, there is no universal format. Rather, companies, journals, and courses select formats to serve their particular audiences, purposes, and occasions.

If the young scientist or engineer—and mature professionals would do well to remember this—learns how to recognize matters of style in his or her writing, he or she will have more control over his or her writing. The way a paper is read will be a result of choices the writer has made in terms of the style. If the choices are deliberate (as they should be), the writer will have more control over how the reader reacts to the concepts and arguments presented.

TABLE 6.1
Choices in Format for Heading Hierarchy and Reference Listings

Format Issue	Purpose	Options
Heading Hierarchy	To rank information	Hierarchy by type size (18 points, 14 points, 12 points)
		Hierarchy by white space (3 spaces, 2 spaces, 1 space)
		Hierarchy by type style (**boldface**, ***boldface italics***, *italics*)
		Hierarchy by number (2.0, 2.0.1, 2.0.1.1)
Reference Listings	To give credit to sources	[Author, Year]: [Jones, 1998]
		[Numbered]: [1]
		[Abbr. author, abbr. year]: [JON, 98]

However, style (itself a combination of several choices by the writer) is a subjective choice made (sometimes unknowingly) by the writer. Different writers and hence different readers have varying ideas about what constitutes good writing style, and so do different organizations (academic, government, and commercial).

For example, passive voice is generally more acceptable in the scientific and engineering disciplines than in the humanities. Furthermore, some issues that readers identify as writing problems may technically be grammatically correct. A sentence can be wordy and still pass all the rules in the grammar handbooks. This fact may make it more difficult for the novice scientist or engineer to determine what is incorrect in the work.

For this reason, some journal editors insist that the reviewers do not apply themselves to grammatical errors (unless they make the document unreadable and/or interfere with the technical argument insofar as the concept needs to be expressed more directly, more elegantly, and more persuasively) and leave such issues to the copy editor or the production editor, thereby assuring consistency throughout the journal.

While the format for many styles of writing is often simple—double spaced and one side of the paper only—in science and engineering works, the format is much more detailed. One reason is to make the reading process efficient as, for example, in a laboratory report when all of the information follows a specific sequence it is easier for readers to locate specific information such as the results. In another instance, having all information in proposals placed under specified headings helps reviewers make fair comparisons of the information in different proposals.

However, no universal formats exist in science and engineering—the format is dependent upon the company, the university, or the journal to which a paper is to be submitted. In short, there are formats for formal reports (laboratory reports and progress reports), and any other documents (proposals, instructions, journal articles, and presentation visuals) that may be written.

In fact, if a scientist or engineer was to pick up several scientific and engineering journals and look at the way headings were done or references were handled, he or she might very well find different formats for each journal. Why are formats in engineering and science so varied? One reason is identity. To a similar extent, companies and laboratories often want their own *report character and style*.

Some common differences found in formats are (1) the hierarchy of headings and (2) the list of the references that are cited in the text.

One reason that a format specifies a hierarchy for headings is so that readers can understand what information in the document is primary and what information is subordinate. The actual ways to represent these hierarchies vary considerably, such as different type sizes for the headings, different amounts of white space surrounding the headings, different typestyles for the headings, and numbering schemes for different order headings. In still other cases, such as the default option of a word processor (header 1, header 2, header 3), the formats call for combinations of these variables.

Likewise, the formats for assigning credit to sources vary considerably. Some formats call for an author-year listing in the text, others for a numbered listing, and still others for an abbreviated listing. Each of these listing systems refers to a section, often named *References* (or, in some cases, *Bibliography*) where readers can find a full citation for the source. The manners in which the full citations are written also vary widely.

Given the wide variety of format issues and the even wider variety of options for those issues, generalized format guidelines cannot possibly present every format option that a scientist or engineer will encounter in science and engineering writing. Such a collection would be cumbersome and, in the end, not particularly helpful. What is important is not that a scientist or engineer learns every format that exists, but that he or she realizes that a specified format will often exist and that format should be followed and contributed to the acceptance of any written work. For those situations in which no specified format exists, a professional format should be chosen that is appropriate for the situation.

An important, but sometimes hazy, distinction in scientific and engineering writing is between format (the way the type is placed on the page) and style (the way that a thought or concept is expressed in words and images).

Generally, style comprises structure, language, and illustration (which all contribute to format). Many journal guidelines (usually available from the journal as *Instructions for Authors*, or a similar title) do not attempt to address the many questions of style; rather, these guidelines focus on stylistic points particular to writing a technical document. The use of style in scientific and engineering writing is often a matter for the writer to address as a personal issue (Strunk, 1918; Alley, 1996).

Writing is an essential skill for the successful scientist or engineer. A scientist or engineer cannot treat writing in the same way that he or she would treat petroleum chemistry or fluid mechanics. Scientific and engineering writing is not a science—it is a craft or even an art. The mental approach to writing differs dramatically from those of making classroom notes as an undergraduate. There is only one reader for such notes (the writer) and conventions used may not be in use by any of the classmates. The same is true for the differences between the solution of technical problems (mental issues) and the communication of those solutions (physical issues—i.e., writing).

6.3   CLEAR, CONCISE, AND UNDERSTANDABLE

It is an unfortunate misconception that scientific or engineering writing (especially in academia) means using long-winded sentences. The obvious result is that the work may not be understood by the audience and the point of the work will be lost for all time.

Simplicity, clarity, and understandable writing are appreciated by all audiences and the document should always follow principles of clear, concise, and understandable communication.

As stated above, the goal in scientific and enquiring writing is not to *sound* intelligent, but to get the concept or intelligent point across while using an acceptable style of writing (Williams, 1994). Each sentence should be as direct and simple as possible. The scientist or engineer should make it easy for the readers to understand. Awkward writing takes the readers away from the points of the paper (or report) that the scientist or engineer is working hard to put across.

Many scientific and engineering students graduate after having read complicated textbooks (many of which are not at all well written or understandable). These types of books do not make sense to the students but are *required* to be used for the course by the lecturer/professor; these books have often been written by the lecturer/professor, who suggests that purchase of the books is necessary for a passing grade (surely a conflict of interest). When the student has to write an assignment paper, he or she may try to imitate this type of writing, often attempting to imitate the style found in the book and missing the most important goal—communicating and being understood. This form of *training* then follows into professional life where (in a university) it may be propagated or (in a commercial organization) it may be changed to match existing concepts of format and style.

If the reader cannot understand what the writer is attempting to *say*, the writer may lose whatever form of credit is available. It is always to be remembered that the most important goal in every scientific or engineering paper is to get the point across as clearly and concisely as possible.

The reader must not be placed in the position of having to read the mind of the writer. If the reader cannot understand the point that the writer is trying to make, interest will wane and the most important goal of getting the point across in as stratight-forward a manner as possible will not be realized.

Thus, the purpose of writing a research paper is not only to present results but also to explain, interpret, predict, suggest, hypothesize, and even speculate. The main purpose of the discussion is to provide a forum in which the author seeks to convince the reader of (1) the reason for the work, (2) the validity of the experimental work, (3) the soundness of the results, and (4) the validity of the conclusions.

At every step, it should be clear to the reader whether the discussion merely interprets results and predicts further outcomes, or launches into more far-fetched speculations. References are essential for this process, but readers are easily annoyed if they are dragged through every publication that has a bearing on the main theme. For the most part, readers expect a coherent interpretation of the results and a demonstration of their relevance. If the discussion must perform intellectual or literary acrobatics to interpret and convince the reader(s), the results are obviously not sufficiently convincing on their own. If reviewers and editors feel this way, they might require additional experiments before accepting the paper.

The writing style is the manner in which a scientist or engineer chooses the strategy to address an issue and an audience. Even though scientific and engineering journals require a specific format, the writing style reveals the writer's personality and also shows that the scientist or engineer understands the audience to which the article is directed. The writing style reveals the choices the writer makes in syntactical structures, diction, and figures of thought. Similar questions of style exist in the choices of expressive possibilities in speech.

The position of a scientific or engineering writer and his or her concept of the audience impose style constraints on the writing. Technical writing, for example, should avoid figures of speech and clichés—precise descriptions are preferred to colloquial terms that might be found more often in more familiar forms of writing.

A scientist or engineer can combine personal style that places him or her at the center of the narrative with the expectations of the audience, but elements of personal style should be used infrequently or not at all in some writing situations. A technical paper that relies on a first person narrator, for example, may cause the reader to question its seriousness or objectivity.

Within the rules of grammar, the scientific or engineering author can arrange words in many ways. A sentence may state the main proposition first and then modify it; or it may contain language to prepare the reader before stating the main proposition. In addition, varying the style may avoid monotony. However, in technical writing, using different styles to make two similar utterances makes the reader ask whether the use of different styles was intended to carry additional meaning.

The scientist and engineer must recognize that the position of the words in a sentence is the principal means of showing their relationship. The writer must therefore, so far as possible, bring together the words, and groups of words, that are related in thought, and keep apart those that are not so related (Strunk and White, 1959).

For a text to be considered science or engineering, it must be factually accurate and written with attention to style and technique. The primary goal of scientific and engineering writing is to communicate information that is defined by established conventions. The information must rely upon documentable subject matter chosen from the real world as opposed to invented from the writer's mind. Another characteristic is that exhaustive research must have been performed to allow scientists and engineers unique perspectives on their fields of study and also permit them to establish the credibility of their work through verifiable references in their texts.

In fact, scientific and engineering writing is a literary prose style in which verifiable subject matter and exhaustive research guarantee the goal of the written work.

6.4   GETTING TO THE POINT

The purpose of a scientific or engineering paper is to communicate results and analysis to the wider scientific and engineering community. The better a paper is written, the more readers it will attract and the more citations it is likely to receive. This alone should be sufficient to convince any scientist or engineer to put significant effort into his or her writing. Unfortunately, this is rarely the case.

There are two kinds of scientific and engineering writing: that which is intended to be read, and that which is intended merely to be cited. The latter tends to be infected by an overblown and pompous style. The disease is ubiquitous, but often undiagnosed, with the result that infection spreads to writing of the first type (Gregory, 1992). Nevertheless, the bulk of scientific and engineering literature is still almost unreadable, and is usually only read by scientists or engineers with a vested interest in the subject. Those who want to read about science and engineering for pleasure are instead advised to pick up the technical pages of a newspaper or a popular-science magazine.

Scientists and engineers should not complain that they lack guidance: there is an abundance of literature on how to write clearly and understandably to attract the interest of the readers (Zinsser, 1976; Strunk and White, 1959). Such books might not cater explicitly to scientific and engineering writing but they are nevertheless valuable as they explain how to organize material in a coherent way, and how to write a manuscript that is both informative and readable.

More importantly, such books convey an important message: authors should write not for themselves but for their readers. Many scientists and engineers would do well to heed this advice, as a clear and understandable manuscript is more likely not only to draw citations but also to be accepted for publication in the first place. Unfortunately, the scientific and engineering literature is still abundant with lengthy, unclear prose that is likely to confuse readers, even those who are familiar with the subject material (Choudhury et al., 2006).

Of course, there are limitations on the style and format of a scientific manuscript. In addition to taking into account the specific requirements of scientific and engineering journals, a paper must generally have an introduction, separate sections on methods and results, and a discussion of the results in relation to the original hypothesis. The very nature of a scientific or engineering paper—presenting and discussing results in an unbiased way—also poses restrictions on the writing style: the passive voice is ubiquitous in order to appear impersonal, and the need to cite relevant references can interrupt the concise and clear flow of text.

However, these rules are flexible enough to allow a paper to be written in both an informative and an interesting way.

Generally, *form follows function* and many scientists and engineers believe that there is nothing more important than their results when, in fact, neither the results nor the paper itself is of utmost importance to the scientific world.

The primary function of a scientific paper is to transmit a message—to convince the reader and the community that this is important research. It is therefore a good strategy to first think about the message before sitting down to write.

Even before the title or first sentence is written, it is helpful to scrutinize the results carefully, as modern science and engineering deals less with demonstrating facts and more with interpreting and discussing results (Horton, 1995). Although a writer's confidence in the results might be fully justified, the ease of his or her convictions sometimes is not.

A scientist or engineer might hold the results as self-evident truths that require no further explanation, but the reader might think otherwise; in fact, many queries raised by peer reviewers are rooted in the writer's assumption that what has become obvious to them through long contemplation and discussions with colleagues will automatically be obvious to the reader. This also means that experimental results should be excluded if they do not contribute significantly to the main message of the manuscript, regardless of how interesting they are. If discarded results are sufficiently substantial, they might form the basis of another paper.

The division of a paper into introduction, methods, results and discussion reflects Aristotle's requirements for introduction, narration, proof and epilogue in oratory—the art of communication using rhetorical skills (Aristotle, 1991).

Although a scientific or engineering paper should not be oratory in the original sense of the word, it can accommodate some rhetoric without compromising its integrity; indeed, rhetoric is an ingredient of good scientific writing. In addition to a simple presentation of the facts, the best writing—scientific or engineering—tries to convince the reader of something (Reese, 1999). In making an argument, the polished scientific or engineering author relies on rhetoric, or the facility of using spoken words or literary composition effectively. Objectivity is the basis of research, but effective communication of science requires some subjectivity on the part of the writer.

However, there is a big difference between subjective—and sometimes even emotional—assessment and scientific rhetoric. Unless the reason for this particular interest is explained, it remains an empty appeal or, worse, an admission that the author does not fully understand the implications of that result. Similarly, subjectivity does not require the pervasive use of adjectives and adverbs to state that a result is *very interesting, highly significant,* or *particularly relevant.*

Readers expect to find certain types of information in particular locations in a scientific paper. Although the divisions between the sections are not set in stone, disregarding them results in a shapeless paper. Excessive experimental details in the results section or unwarranted reiteration of results in the discussion will leave the reader wondering what the main message is.

A result must be presented before it can be discussed, and any results that do not add to the point being discussed should be excluded. It is possible to cross the internal divisions of a paper to enhance the message and to preserve the flow of arguments, but this should be done judiciously.

The title is the single most important phrase in the entire paper and the impact of the title must not be underestimated—a reader who cannot extract the significance of a paper from its title is unlikely to read further. Long titles can be more informative, but they are less likely to catch the attention of readers who scan quickly through journal contents or article listings. Short titles can be more attractive but they carry the risk of being too cryptic. Titles using puns or clever wordplay, although not necessarily informative, can attract readers' interest, but this should not be done at the expense of information that portrays the article's content.

The title should first be informative, and any wordplay should only be used as embellishment. When formulating the title, it is helpful to consider whether or not it sounds sufficiently informative.

Equally important is a good abstract. It is frequently on the merits of the abstract alone that a reader decides whether to peruse a paper. The whole article might be a treasure trove of information, but if this does not come across in the abstract, the article might be ignored. There are two main ways to write an abstract: free-form and structured (Fuat et al., 2003; Gallagher et al., 2003). A free-form abstract is usually written as a single paragraph, whereas structured abstracts are organized into sections, the most basic of which are objectives, methods, results, and conclusions (Hartley and Sydes, 1997). Regardless of any personal view on the best method, writing the first draft of an abstract in a structured form might help to get a better idea of how much of it should be devoted to different aspects of the paper to achieve a well-balanced text. There are also divided views on whether to write the abstract at the outset or as the last step (Baillie, 2004; Fisher, 2005). In any case, it might be a good exercise to try both methods and see which works better.

There are those writers who prefer to write the abstract first followed by the corpus of the paper. This becomes a personal preference although there are benefits and dangers of such an approach (Baillie, 2004).

There is a clear difference between an introduction and a literature review, as the latter is an article type in itself. Consequently, a good introduction should not cover as much of the literature as possible within the space constraints and the main goal is to draw a map of the research area, situate the manuscript within this map, and put its aims, results, and interpretation into context. In general, an introduction moves from a general overview to address specific questions.

A short historical overview could lead to a brief description of the state of current knowledge and highlight any gaps. This provides the roadmap for stating the problem that the paper addresses, its aims, and the results. Of course, this is not a rigid framework, rather a flexible guide that can be changed to suit the aims and purpose of each paper.

The experimental section should be specific and sufficiently detailed to allow other scientists or engineers to reproduce the experiments, but no more. They should be able to use it as a set of clear instructions on how to perform the work. One common mistake in describing methods is failing to provide essential information. It must always be remembered whether the description of a particular experiment is sufficient to repeat it. Procedures adopted from the literature should of course be referenced, but they could also be outlined in brief for the benefit of the reader.

Results should be presented in a coherent and organized way that tells a logical, rather than a chronological, story. Of course, research is rarely a linear process from observation to hypothesis to experimental proof: any scientist knows how often his or her research backtracked or branched off in unexpected directions—there is nothing more disconcerting than trying to assemble a story from a jigsaw puzzle of results.

It is essential to study the results in detail and organize them in a logical fashion before writing the paper; otherwise this disorganization will manifest itself in the paper and be noted by readers and reviewers alike. Furthermore, including extraneous results that provide little support for the main theme will dilute the message and confuse the reader.

In summary, (1) short sentences are the key to an understandable document—a very general rule is that a sentence should not go for more than three typed lines because long sentences can lose or confuse the reader; (2) structure in writing is important and the writer should never underestimate the use of words such as *first*, *second*, and *third* as a means of introducing ideas, (3) there should be only one idea per paragraph—introducing a second idea in a paragraph can lead to confusion in the eyes of the reader, (4) key terms should always be explained—the document or presentation may be written for a specialist audience but there may be a number of nonspecialists who are attempting to learn about the subject matter.

6.5 MINIMIZING VERBAL OVERKILL

Verbal overkill (*wordiness*) is the term used to describe style problems that involve using more words than is absolutely necessary. Such words are often described as *filler words* that do not actually add anything to the meaning of the sentences. In fact, filler words and phrases become more obvious and act as delays in getting the reader to the point of the paper or report. This will eventually lead to audience frustration and failure to assimilate the points of the work.

Any scientific or engineering written communication or verbal presentation that seems to go farther than would be necessary to achieve getting the point across is overkill. Restating the presentation message repeatedly, even using different words or approaches, is verbal overkill. If the communication is clear and concise there is no need to belabor the point.

To avoid such communication problems, the scientific or engineering author must first be clear in his or her own mind at to what the message is. He or she must use words that easily communicate his or her thoughts and the language must be clear and concise (short and simple). The sentences must be easily understood; there should be no misunderstanding by the audience.

Thus, scientific and engineering writing should be concise sentences should contain no unnecessary words and paragraphs should contain no unnecessary sentences (Strunk and White, 1959). This requires not that the writer make all his sentences short, or that he avoid all detail and treat his subjects only in outline, but that every word count. In short, there should be no *verbal overkill*.

Good writing does not need an abundance of adverbs and adjectives. If the presentation of results and the ensuing discussion are logical and conclusive, the reader will be able to follow them more easily than if he or she must traverse a forest of unnecessary words.

A sentence should contain no unnecessary words, and a paragraph no unnecessary sentences, for the same reason that a drawing should have no unnecessary lines. This does not require that the writer makes all his sentences short, or that he avoid all detail and treats his subjects only in outline, but that he makes every word essential (Strunk and White, 1959).

Long-winded sentences with multiple clauses, disclaimers, and parentheses (the excessive use of parentheses) are hard to read and are guaranteed to discourage even the most interested readers (Gage et al., 2006). Good writing involves self-editing to clean up the language until the prose is clear and understandable.

The term *wordiness* is often used to cover style problems that involve using more words than is necessary. When scientists and engineers (especially those of an academic leaning) talk, they use a lot of little *filler* words that do not actually have anything to add to the meaning of the sentences—listening to a professor and waiting for him/her to get to the point can be a new experience in pain and terror!

In writing, these filler words and phrases become more obvious and act as delays in getting the reader the point. If the writer has similar delays in sentences, the readers might get frustrated. The reader may even start skimming the paper, which make it seem that the writer has wasted his or her time in making the effort to communicate with the readers.

Wordiness may derive from problems unrelated to writing style, such as: (1) uncertainty about the topic, (2) lack of a developed argument, or (2) lack of evidence. If one or more of these reasons is the case, the writer must reconsider what he or she needs to say in the document rather than struggle to get to the point. It might be that the writer is using writing as a way to discover his or her point. He or she should first take the time to clarify the point with the reader(s) in mind.

Some writers think that they can improve their style by sounding more academic or by using multisyllabic words. This can be a hazardous mistake of the greatest magnitude.

First, the writer should be sure of the meaning of such words as the use of such words may be dangerous unless the meanings have been double-checked *in a dictionary*. Many times, an inappropriate synonym will make the writer sound like he or she does not know what he or she is talking about or, worse yet, it might give the impression of plagiarized material from a source the writer did not understand.

Wrongful use of such words or the thoughts of plagiarism will more than likely cloud the point rather than clarify it.

The exponential increase of both primary papers and reviews means that scientists and engineers are under increasing pressure to keep up with the literature in their field of interest.

Furthermore, as the performance of text retrieval and analysis algorithms to draw meaningful information from the literature improves, scientists will increasingly rely on these to harvest relevant papers from the deluge of available information. However, scientists will still read papers if they think that the title is interesting or that the message or question being answered is important.

The better the paper is written and the more logical its arguments, the higher the chances that the reader will proceed beyond the abstract and find it convincing enough to cite. Consequently, it is of utmost importance to keep two things in mind throughout the writing process: the main message and the reader. After all, the author's goal is to convince the reader that this is important research. If a paper ignores readers' interests, they in turn might ignore the paper.

A technical paper should address one main question, and the failure to do this is one of the most common reasons for reviewers to reject a manuscript (Lambert et al., 2003).

A good research question should be specific, novel, and of interest to the scientific and engineering communities and will dictate the choice of journal and its readership (Perneger and Hudelson, 2004). Readers of highly specialized journals will be easily turned off by lengthy explanations of what is common knowledge in the field; conversely, readers of general journals might need more background information to be able to follow the arguments. When writing a paper, scientists and engineers should put themselves in the readers' position so that they can assess the depth of detail from their point of view.

Those who have studied the art of writing are in accord on one point insofar as it is the surest way to arouse and hold the attention of the reader: By being specific, definite, and to the point without any interference from superfluous insertions (Strunk and White, 1959).

Finally, wordiness may derive from issues that are unrelated to a scientist or engineer's writing style. For example, if the writer is uncertain what he or she needs to write about the topic or if there is a lack of a developed argument or even lack of evidence, there is the need to reassess the reason for writing the paper or report. In fact, there may have been insufficient prewriting preparation (Chapter 5). Taking the time to organize one's thoughts is always time well spent.

6.6 ETHICS IN WRITING

Although reference has been made to ethics elsewhere in this book (Chapter 3, Chapter 4, and Chapter 8), there is the need to expand on the application of including ethics as part of the writing style. Indeed, both the Modern Language Association and the Chicago Manual of Style advocate the requirement for ethics in writing by dutiful and correct citation of prior work.

Ethical writing, which is itself part of the writing style, is straightforward if the author is properly prepared and *willing* to acknowledge the work of other scientists and engineers.

The writer has to be forthright and honest while working on the document and must have all of the basics outlined so that there is no room for error—intentional or unintentional. The only way a scientist or engineer can be prepared and not make any mistakes is through research, which is an important factor in ethical writing.

Applying the term ethics to writing a technical paper (or any technical document, for that matter), requires application of moral behavior to the writing task (Speight and Foote, 2011).

This requires the scientific or engineering writer to reflect his or her own thoughts and personal understanding. For the inexperienced writer, it may seem strange that he or she would need to determine the moral boundaries or principles himself/herself.

The basic principle is very simple: if similar work has been published at some past time, it should be cited. There should be no excuses such as the work was published in an obscure journal that the writer never read (even though the researcher did read the journal and was hoping that others had not) or the author of the 200-page company report does not cite the open literature because it was not relevant (even though the company researcher used the published data and experimental techniques available through the open literature as the basis for his or her work) (Speight and Foote, 2011). Such excuses are not acceptable.

A different form of unethical behavior in writing is when the author (or presenter) writes (or says) something that can cause the reader to believe something that is not true. This can be done by lying, misrepresenting facts, or *manipulating* data to favor the author's opinion and objectives. Put simply: data facts are data facts and must be represented in that way. Statistically smoothing data (whether this be through taking averages or subjective placement of a line on a graph to calculate the slope of the line) to change the interpretation is not acceptable—figures, charts and tables should be used cautiously to make sure that they are not misleading. Unfortunately, there are instances of scientists and engineers who have stretched the truth to support their own individual points and theories (Speight and Foote, 2011).

While the scientist or engineer may feel that he or she must write or present material persuasively to convey a hypothesis to the audience, it is not wise to attempt to over-persuade the audience. The audience may find out that what was written or presented only benefited the author and not science or engineering.

Another frequently used component of unethical writing is plagiarism, which is the deliberate misrepresentation of the source or facts whereby the writer claims the ideas as his or her own ideas. When a scientist or engineer is researching professional documents for use in a document, he or she should make sure that the source is correctly acknowledged—all sources should be cited and credit given to all of the researchers. In short, all sources should be properly documented so that the reader is not misled.

Thus, in scientific and engineering writing, the writer should avoid data manipulation and misconceptions of the work of others. The author must not use false or skewed data or use such data in an attempt to deceive the audience for his or her own benefit. For example, if a control experiment was omitted (as part of laboratory work or fieldwork), the work should not be published until the control experiment has been performed, even if the results of the control experiment change the outcome of the author's hypothesis.

Above all, the scientist or engineer should avoid using misleading words or phrases or manipulating data, and it is essential that he or she be open to alternative viewpoints.

Thus, in preparing any type of scientific or engineering writing, the author will discover published conflicting viewpoints and should be aware of the hypotheses of other scientists and engineers. Acknowledging and discussing several opinions and ideas in any technical document or presentation will make the author more persuasive and the inclusion of all other hypotheses will remove the perception of bias.

REFERENCES

Alley, M. 1996. *The Craft of Scientific Writing*. 3rd edition. Prentice Hall, Upper Saddle River, New Jersey.

Aristotle. 1991. *The Art of Rhetoric*. Editor: H. Lawson-Tancred. Penguin Books, London, United Kingdom.

Baillie, J. 2004. On writing: write the abstract, and a manuscript will emerge from it! *Endoscop*, 36: 648–650.

CMS. 2010. *Chicago Manual of Style*. 16th edition. University of Chicago Press, Chicago, Illinois.

Choudhury, B., Risley, C.L., Ghani, A.C., Bishop, C.J., Ward, H., Fenton, K.A., Ison, C.A., and Spratt, B.G. 2006. Identification of individuals with gonorrhea within sexual networks: a population-based study. *Lancet*, 368: 139–146.

Fisher, W.E. 2005. Abstract Writing. *J. Surg. Res.*, 128: 162–164.

Fuat, A., Hungin, A.P., and Murphy, J.J. 2003. Barriers to accurate diagnosis and effective management of heart failure in primary care: qualitative study. *British Medical Journal*, 326: 196.

Gage, M.J., Surridge, A.K., Tomkins, J.S., Green, E., Wiskin, L., Bell, D.J., and Hewitt, G.M. 2006. Reduced heterozygosity depresses sperm quality in wild rabbits, *Oryctolagus cuniculus*. *Curr. Biol.*, 16: 612–617.

Gallagher, T.H., Waterman, A.D., Ebers, A.G., Fraser, V.J., and Levinson, W. 2003. Patients' and physicians' attitudes regarding the disclosure of medical errors. *J. Am. Medical Associations*, 289: 1001–1007.

Gibaldi, J. 2009. *MLS Handbook for Writers of Research Papers*. 7th edition. Modern Language Association of America, New York.

Gregory, M.W. 1992. The infectiousness of pompous prose. *Nature*, 360: 11–12.

Hartley, J., and Sydes, M. 1997. Are structured abstracts easier to read than traditional ones? *J. Res. Reading*, 20: 122–136.

Horton, R. 1995. The rhetoric of research. *British Medical Journal*, 310: 985–987.

Lambert, V.A., Lambert, C.E., and Tsukahara, M. 2003. Basic tips about writing a scholarly manuscript. *Nurs. Health Sci.*, 5: 1–2.

Perneger, T.V., and Hudelson, P.M. 2004. Writing a research article: advice to beginners. *Int. J. Qual. Health Care*, 16: 191–192.

Reese, D.M. 1999. Restoring the literacy to medical writing. *Lancet*, 353: 585–586.

Speight, J.G., and Foote, R. 2011. *Ethics in Science and Engineering*. Scrivener Publishing, Salem, Massachusetts.

Strunk, W. 1918. *Elements of Style*. W.F. Humphrey Press, Geneva, New York.

Strunk, W., Jr., and White, E.B. 1959. *The Elements of Style*. Macmillan, London, United Kingdom.

Williams, J.M. 1994. *Style: Ten Lessons in Clarity & Grace*. 4th edition. HarperCollins, New York.

Zinsser, W. 1976. *On Writing Well: An Informal Guide to Writing Nonfiction*. Harper & Row, New York.

7 Teamwork

7.1 INTRODUCTION

A *team* (within the context of this book) is a group of scientists and engineers working toward a common goal for the good of the team and the organization. On the other hand, in the *group* scenario, the members work independently and they often are not working toward the same goal.

The concept of teamwork is extremely important to the success of any team of scientists and/or engineers. Teamwork and unselfishness create the backbone of a great team; without them a team cannot realistically compete. A group of *prima donna* scientists or engineers may appear to be unbeatable in theory and on paper, but if the individual members do not work well as one unit, the chances of success are limited. Team members (somewhat less than *prima donnas*) working as one cohesive unit are the key to success.

While many scientists and engineers may consider themselves *loners*, there are advantages to teamwork. For example, it is rare that a single writer tackle a complex writing project, which requires the skills, talents, and insights of several writers working collaboratively. In addition, writing projects must be completed within a specified time period, which forces organizations to employ more than one writer to complete each project. Finally, a group of writers working on a project instead of a single individual ensures that the project will still be completed successfully even if one of the members of the group is incapacitated (or shirks his or her duty).

When scientists and engineers work in teams, there are two separate issues: (1) the goal and the problems involved in reaching the goal and (2) the mechanisms by which the team acts as a unit and not as an uncoordinated group.

As members of a team, each scientist or engineer usually places his or her individual interests and opinions second to the unity and efficiency of the team. Although the individual member of the team is still important, effective and efficient teamwork must reach beyond the accomplishments of the individual—the most effective teamwork is produced when all the individuals involved harmonize their contributions and work toward the common goal.

Scientific and engineering projects require that the members of a team work together to achieve a common goal and typically involve assigning each team member a specific task that he is responsible for completing, which helps to develop trust within the team. Furthermore, teamwork is an important aspect of presenting the results of multiperson scientific and engineering research to a wide audience.

When scientists and engineers work in groups, there are two separate issues to be addressed: (1) the task and the problems involved in completing the work and (2) the process by which the group acts as a unit and not as a loose collection of individuals. Having a group of technical people working on a project instead of just one individual is often considered to ensure that the project will be completed successfully even if one of the members of the group is incapacitated. But this is not always the case!

The concept of teamwork is extremely important to the success of most scientific and engineering projects (Bell, 1995). Teamwork and unselfishness create the backbone of a great team; without them a team cannot realistically operate. A team consisting of a group of experienced and well-published scientists and engineers will not be successful if they do not work well as one unit. The team working as one cohesive unit is always the key to success.

A productive and successful technical team has members that share common goals and have some level of interdependence that requires both verbal and physical interaction. Teams come into existence through shared attitudes about a particular area of science or engineering. The members may be brought together (by management) for a number of different reasons, but their goals are the same—to achieve success through peak performance. Every member of the team is accountable when it comes to teamwork and no one member should be given praise or blame. The old adage written by Alexandre Dumas in his 1844 novel *The Three Musketeers*—All for one, one for all—is operative!

Good teamwork involves mutual accountability and togetherness among members of a well-knit team. To make the writing project a success every team member needs to combine their efforts to increase what the team can accomplish. If individual team members master the fundamentals and work together as one unit the team will be successful—the whole is greater than the sum of the parts!

Teamwork is something that must be a high priority and given constant attention. Every team member needs to understand how important it is for them to work smoothly together if they want to be successful. Each scientist or engineer must be dedicated to the whole team and be willing to act unselfishly. When challenges arise (as they always do), the team needs to have the resources, accountability, and commitment to deal with them in a constructive and positive manner. A sense of teamwork will play an integral part in this.

Thus, scientific and engineering writing projects require team members to work together to achieve a common goal. Like all projects, writing projects typically involve assigning each team member a specific task that he is responsible for completing, which helps to develop trust within the team.

In the scientific and engineering context, methods may be used to provide a measure of the benefits of teamwork, which are useful for monitoring the progress (or lack thereof) of a writing project. Responsible scientists and engineers often prove to be successful. The sense of responsibility brings awareness about their individual duties.

Teamwork is essential for competing in the modern scientific and engineering global arena, where individual perfection is not as desirable as a high level of collective performance. In scientific and engineering enterprises, teams are the norm rather than the exception. A critical feature of these teams is that they have a significant degree of empowerment, or decision-making authority.

Teamwork is an important part of the scientific and engineering working culture and many organizations now consider teamwork skills when evaluating a person for employment. Most organizations (universities are often the exception) because of the professor-student relationship rather than the department-professor relationships) realize that teamwork is important because either the product is sufficiently complex that it requires a team with multiple technical and administrative skills to produce, and/or a better product will result when a team approach is taken. Therefore, it is important that young scientists and engineers learn to function in a team environment so that they will have teamwork skills to benefit the workforce.

Universities would be well advised to include courses on collaborative learning as part of the degree programs. There is the need to develop good scientific and engineering team exercises, and this is where understanding of the interactions of the team members can be of considerable benefit to the young scientists and engineers.

7.2 EFFECTIVE TEAMWORK

Effective teamwork can produce extraordinary results but effective teamwork does not happen automatically. In fact, there are several characteristics that guide members to effective teamwork in general (Table 7.1) (Larson and LaFasto, 1989) and which can be applied to a team-oriented writing project. These characteristics are, however, not *cast in concrete* as hard-and-fast rules but can (and will) vary with the nature of the writing project and the nature of the organization.

TABLE 7.1
Characteristics of Effective Teams

The team must have a clear goal:
The specific goal objective must be expressed concisely so that all members of the team know when the goal has been met.

The team must have a results-driven structure:
Team structure is often difficult to define but should be allowed to develop based on (writing) expertise and results.

Team members must be competent:
The Peter Principle* must not operate and each member of the team must work at his or her given level of expertise or competence.

The team must have unified commitment:
All team members, while not agreeing on every issue, should direct their efforts toward the team goal.

Collaboration is essential:
There must be a climate of trust among the team members. Each member of the team must know what is expected of him/her both individually and collectively.

There must be principled team leadership:
The team leader must have good leadership (and writing) skills and exhibit behavior that is beyond reproach. He or she must not put himself/herself above the team in an attempt to achieve personal recognition or benefit from the position—the team should serve the team and consider the first priority to be the collective interests of the team.

* Peter, L.J., and Hull, R. 1969.

However, for the most part, and in the context of this book, a team is brought together to produce a written document, report, or presentation. Team growth, and professional growth of the individual members, will benefit from teamwork.

Foremost, at some stage of the proceedings, a team leader (principal author) should be designated either by seniority or by general agreement among the team. In a company setting the team leader may have been designated by upper management. After this, the team must decide on a clear goal (i.e., the production of a technical document) that calls for a specific performance objective, which is expressed so concisely that everyone knows when the objective has been met. In the case of a paper for publication or a report for internal company circulation, the team should be allowed to operate in a manner that produces results. It is often best to allow the team to develop the structure.

In many cases, a single writer may not be able to tackle a writing project that requires the skills, talents, and insights of several writers working collaboratively. Thus, the team leader may not be the best writer but, short of a mutiny, the team members should support the designated leader and move toward achieving the goal in the best way possible.

In such a case, the team members will not always agree on everything but all individual members of the team must direct their individual (even collective) efforts toward the designated goal. If it is perceived that the efforts of individual members of the team are directed toward personal goals, then the team will confront this and resolve the problem. To this end, in any setting, the team must have some competent members so that they can tackle the writing task at their respective levels of scientific or engineering knowledge.

Irrespective of differences of opinion about the team leader of the technical matter under consideration for the document, there must be a unified commitment and the team must have a collaborative climate (Morgan and Murray, 1991). Moreover, a climate of trust must be evident from honest, open, consistent, and respectful behavior. For example, if the team leader (or any member of the team) is seen as being somewhat out of his or her depth, the team members should help him/her. Without a climate of trust and/or the willingness to help each other, the team will fail and an inferior written product will be the result. To this end, encouragement and praise (both internally and externally) works just as well in motivating teams as it does with individuals.

In addition, the team must work using the highest ethical standards (Chapter 6). Team members must know what is expected of them individually and collectively. Vague statements such as *positive attitude* and *demonstrated effort* are futile expressions when hard work and ethics are at stake.

In order to ensure the highest ethical standards, the team must have principled leadership, whether or not the team leader is the best writer in the team. Team members must know that the team leader has the position because he or she has good leadership skills and (irrespective of the writing abilities) is working for the good of the team. Obviously, team members will be less supportive if they feel that the team leader is putting himself/herself above the team, achieving personal recognition, or otherwise benefiting from the position.

Thus, when a team is being formed, the members (especially the young scientist and engineer) must be aware of the reason why they were included in the team and, more important, the boundaries of acceptable group behavior. The members must establish their respective positions within the team and seek the leader's guidance.

At some stage shortly after the formation of the team, the novice scientist or engineer may become impatient about the lack of progress. He or she may even disagree with the actions they should take, or that are being taken by other members, because he or she is faced with scientific or engineering ideas that are unfamiliar and actually move the scientist or engineer outside of his or her comfort zones.

As the team evolves, team members accept the team and begin to reconcile differences. Any emotional conflict that developed during the initial stages of the team formation may even be reduced as relationships become more cooperative. As a result, the team is able to concentrate more on the work at hand (i.e., the production of a technical document) and start to make significant progress.

And then when the team members have learned of each other's strengths and weaknesses, and learned what their roles are, progress is made. Members are open and trusting and many good ideas are produced because they are not afraid to offer ideas and suggestions. They are comfortable using decision-making tools to evaluate the ideas, prioritize tasks, and solve problems. As a result, much is accomplished and team satisfaction and loyalty are high.

The team has become a team and the project goal will be realized, whether it is a paper for publication, a report for internal company circulation, or any other form of written document.

It is imperative for team members to realize that teams do not just form and immediately start working together to accomplish great things. There are actually stages of team growth and teams must be given time to work through the stages and become effective.

Furthermore, once the team has been assembled and before the writing begins, the team must collectively identify the audience (or client) (Chapter 4). The team should (1) analyze the overall writing project, (2) conceptualize the work to be produced, (3) create a broad outline of the document, (4) divide the document into segments, and when necessary (5) assign each segment to individual team members, often on the basis of expertise. In the planning stage, the team projects a schedule and sets any writing style standards that team members are expected to follow.

Brainstorming, a necessary precursor to any team writing, is the valuable collective activity that provides insight into the project. Effective brainstorming in teams, both in discussion and on paper, leads to the creation and implementation of creative ideas that move a project forward. Early brainstorming as the team formulates the writing goals can help avert *gang thinking*, where someone (usually the most senior member of the team) suggests an idea and everyone agrees because they are not really thinking of alternative possibilities.

In summary, when a written document or a slide presentation is a group effort, the primary contributors should be identified and their roles defined. Each team member knows the level of detail required, the audience, and the deadline. In addition, each contributor should be given an outline of the entire document, so that everyone knows how their section fits into the whole.

7.3   SHARING AND ASSIGNING RESPONSIBILITIES

The goal of science and engineering is to put all the pieces of the puzzle together from different viewpoints, thus finding the answers. Without teamwork, progress in science and engineering would be much slower and sometimes at a standstill with regard to theory and analysis of a perplexing problem or question. Every scientist or engineer has a talent or perspective of knowledge of which others are unaware or which they do not equally share. When many come together as a team to accomplish one aim or goal, questions are answered, the individual talents of the team members are utilized in specific areas relative to the subject matter, and solutions to perplexing problems are solved.

In science and engineering, not all *groups* are teams, but all teams are groups. The difference between a team and a group is that the former is interdependent for overall performance while the latter only qualifies as a team if its members focus on helping one another to accomplish organizational objectives. As a result, teams have emerged as a requirement for scientific and engineering success.

In fact, it is a common problem that only one or several members of the team complete the bulk of the work while the rest of the group members function as spectators. To prevent such a situation, the team members must be aware of this common problem, group members must assess the strengths, talents, and weaknesses of all team members, and tasks must be assigned accordingly.

Before any collaboration begins, all members of the team must understand and agree to their respective responsibilities. The duties of each team member must be defined at the beginning and any concerns or potential problems must also be discussed. Open communication at the beginning of the process can avert serious problems once the team members are involved in the project.

It is necessary to be responsible while working as part of a team in an organization in which work quality is very important. Sharing responsibility is crucial as the tasks are shared among team members, each of whom is responsible for delivering flawless work. Higher authorities need to be careful that no member of the team is overburdened by workload.

If one team member is overburdened simply because he or she is an expert at completing tasks flawlessly, overwork will eventually affect her/his productivity and he or she will not be able to focus on certain tasks and apply his or her skills efficiently, due to paucity of time. Another danger of such a practice is that remaining team members become complacent and lethargic, largely resulting in reduction of the team's potential. Therefore, to avoid the possible loss, each team member must be assigned a reasonable workload based on his expertise and area of specialization.

The practice of recognizing the skill set possessed by each member of the team and allocating tasks accordingly is largely a duty of the team leader. If every team member recognizes his responsibility toward the work, the team would succeed in its endeavors. In the event that an important task is assigned, tasks need to be shared among team members with a commitment to achieve the best results.

The approach to building an effective team of scientists and/or engineers and deciding what they are to do is a subjective question that cannot be resolved by applying any one pattern. There are many variables involved in analyzing and arriving at an appropriate plan in which to implement a strategy.

Certain basic elements must be implemented to create an efficient and workable team concept:

- Team members need to feel secure with their genuine and honest participation in implementing the team concept.
- Workers need to honestly believe that the rules apply equally to team members and managers. Managers cannot reap huge benefits (by *passing the buck* for a mistake) while team workers suffer.
- Each team member must feel that her/his input is important and is taken seriously. There is nothing more discouraging than a team member coming up with an excellent idea and being ignored.
- Team members must (be made to) feel equal to one another and should elect the team leader and not have the company appoint one.

These are basic elements of what must be present in order to develop a good team concept. Other elements are industry and business specific.

When a writing team in a dynamic environment splits into a number of subteams, each working on a different feature or bundle of features, this is a challenge for scientific and engineering authors. If the subteams work in parallel, all of the features of the potential publication may simultaneously reach the stage where they can start combining all of the efforts. That may usually be quite close (too close) to the submittal or presentation release date and this can cause major stress for the team and can require frantic last-minute efforts to complete the work in a timely manner.

Team members depend on each other to achieve a common goal. Each brings a specific skill to the problem and the application of their individual skills leads to solutions of perplexing problems. The members must be responsible for the flawless and timely conclusion of their individual contributions, feel equal to each other, and feel secure in their trust for each other.

When two or more scientists and engineers work on one writing project, they are all responsible for the project's shape and success but, like the workload, the degree of responsibility isn't always equal. One of the team may be responsible for the actual writing, while others are responsible for fact-checking and making sure the document achieves the goals that were established at the onset of the work. Within those broad areas of responsibility are smaller but equally important obligations.

For example, if one team member is to take the work of the team and produce a readable document, the other team members are responsible for supplying the author/writer with the necessary information. Furthermore, the other (nonwriter) team members should be prepared to be available to the writer when necessary to answer questions or clarify material. The (nonwriting) team members may also need to allow the writer access to additional documents or sources to supplement the information to ensure that all necessary works are cited in the final document. In fact, it is the responsibility of all team members to ensure that acknowledgment of the work of others (in the form of citations) has been done correctly and there are no omissions (see below).

The writer in this scenario must make sure that he or she has all of the necessary material to write the manuscript. If pieces are missing, the writer is responsible for advising the relevant team members what is missing and what is needed. When it comes to the actual writing, the writer is obligated to give the other team members sufficient time to review the manuscript and, if necessary, edit or correct the work.

One of the most important aspects of scientific and engineering writing is making use of the ideas of other scientists and engineers. This is important as the scientific and engineering author needs to show that he or she has understood the materials and that the ideas of other authors are usable. In fact, this is an essential skill for every scientist and engineer. For many scientists and engineers, it may be a difficult and/or complex activity to write from other published work (Spack, 1988). It is an option that when this form of citation is used the writer can (some would say *should*) use his or her own words, unless a direct quote is being used. Above all, the author must make it obvious when the words or ideas that he or she is using are taken from another author (by using the appropriate citation). A scientist or engineer must not use another author's words or ideas as if they were his or her own: this is plagiarism, which is regarded as a very serious offense (Speight and Foote, 2011).

In fact, it is the responsibility of all team members to ensure that acknowledgment of the work of others (in the form of citations) has been done correctly and there are no omissions.

The team members, as well as the individual scientific or engineering author, must remember that the object of scientific and engineering writing is for the author to publish (or report) his or her own work and ideas on a particular subject and to present his or her ideas that have evolved as the work progressed. The scientist or engineer can do this by reporting the works of others in his or her own words by paraphrasing, summarizing, or synthesizing if there is the need to use information from several sources (with acknowledgment or citation). In all cases the writer *must* acknowledge the work of other scientists and engineers.

Briefly, *paraphrasing* occurs when an author writes the ideas of another author in his or her own words, and it is useful when the author wishes to use (and acknowledge) the work of others to support his or her own view or theory. During paraphrasing, a writer may/should change the words and the structure of the cited work but should keep the meaning the same.

On the other hand, *summarizing* another author's work involves producing a shortened version of a text, which will contain the main points in the text but is written in the author's own words. Summarizing is a blend of reducing a long text to a short text and, at the same time, selecting relevant information, which is useful when a scientist or engineer is using the work of others to support his or her own view or hypothesis. In addition, a well-written summary shows that the writer has understood the work of the previous author(s).

Synthesizing is a combination, usually a shortened version, of several published works into one or more paragraphs, which contain the important points from those works and is written in the author's own words. To include a synthesis, the author needs to find suitable sources and then to select the relevant parts in those sources for inclusion in his or her document. However, the information from all of the sources should fit together into one continuous text. Again, proper acknowledgment of the source is essential.

As already noted (Chapter 6), there should be no excuses for not acknowledging prior and relevant work. Excuses such as the work was published in an obscure journal that the researcher never read (even though the researcher did read the journal and was hoping that others had not) or the author of the 200-page company report does not cite the open literature because it was not relevant (even though the company researcher used the published data and experimental techniques available through the open literature as the basis for his or her work) (Speight and Foote, 2011). Such excuses are not acceptable.

7.4 CREATING A MANUSCRIPT

First and foremost, before the writing begins, the team should collectively identify the audience (or client) (Chapter 4), the purpose and goals, the scope of the project, and the deliverables (documents, presentations, artifacts, or products).

When the deliverable is a document, the team must analyze the overall project, create a broad outline of the document, divide the document into segments, and assign each segment to individual team members, often on the basis of expertise. In the planning stage, the team projects a schedule and sets any writing style standards that team members are expected to follow (Chapter 3 and Chapter 5).

Integrated documents are typically complex and, more often than not, include flaws that cause them to be less than perfect—if such a document is sent for publication (in the open literature or as an internal company report) it will not be successful. Rejection by the journal reviewers or by company management is the usual fate of such a document, to the detriment of the coauthors.

A rejected document may result from: (1) missing information, (2) illogical or nonfunctional layout (Chapter 3), (3) inconsistencies in point of view or treatment of the subject when the authors cannot agree on their hypotheses or the meaning of the data, and (4) unsatisfactory recognition or omitted citations of the published work by prior authors (Chapter 3 and Chapter 5).

Invariably, all documents contain both strong points and weak points but when the weak points outnumber the strong points (and they have not been corrected), the document is unacceptable and the authors are classed as failing in their task of writing an acceptable document.

To avoid such possibilities, collaborative scientific or engineering writers must actively scrutinize their documents collectively and depend upon one another to employ objective and constructive criticism. Criticism on the basis of differing personalities within the team is not acceptable.

It is also worthy of note that collaborative writing often lacks the means to create a clear and precise document—each writer may have a different image of the final work. A team of individual writers may use concepts and terms that are imprecise and deficient for communication among team members. To mitigate this problem, it is useful for a team of writers to meet and express verbal representations of the finished product. In addition, they may create outlines, prototypes, and partial rough drafts, and by reviewing portions of a document throughout the different stages of its development, team writers can become more certain of reaching a consensus about the meaning they wish to convey in the final draft.

The early appointment of a team leader (see below) who can resolve such conflicts (without being too dogmatic in his or her approach) will guide the team though such difficult periods.

The process of creating and preparing a manuscript for publication in the open literature or as an internal company report is a complex process when performed by an individual and becomes even more complex when undertaken by a team. The (individual) scientific or engineering writer must maintain control of the document theme, the organizational structure, the level of detail, and all other aspects from beginning to end.

Thus, to create an acceptable manuscript, all members of the writing team must have a collective image of the desired outcome. Tasks must be assigned to each individual and given a comprehensive schedule. Development of a schedule is always valuable, especially when those team members whose task is dependent upon the completed work of others will receive such work in a timely manner. In addition, individual work will more than likely need to be revised to facilitate the collective image, as well as to correct individual errors and insufficiencies.

For the individual scientist or engineer, a strong commitment to his or her own personal style of writing can create difficulties in collaborative writing projects. Team members who consider their individual component of a collaborative work as a private document and assume the prerogatives of *ownership* of the piece will not only jeopardize the group work but will also provoke conflict within the group.

Those who are strongly committed to high quality work should find ways to focus this attitude on efforts to facilitate teamwork and pride in a successful collaborative outcome.

Again, the early appointment of a team leader (see below) who can resolve such conflicts will guide the team though such difficult periods.

It is convenient to acknowledge that every document ever written is reworkable and subject to revision. Having made that statement, it is certain that there are documents written since time immemorial that have not been reworked or rewritten. But to accept the premise is the essence of scientific and engineering writing.

Revision is an important component in all forms of scientific and engineering writing. In the revision process, all team members see document changes suggested by others and changes should be discussed by the team before becoming part of the next draft of the document.

Moreover, team members should be encouraged to send comments, questions, and data to one another during the writing process. This will allow team members to develop and suggest alternative patterns of organization, types of graphics, and tentative drafts.

However, team members should be discouraged from excessive reworking. Such obsessive compulsive behavior (while perhaps uncommon when a team leader/writer was appointed early in the writing project) may come from a team member who has played a limited role in the project but needs to attempt to establish his or her importance. Such behavior will only serve to create animosity within the team. Again, the team leader (writer) needs to step in and resolve such an issue.

Finally, when creating a document as a team effort (indeed also as an individual effort) it is difficult to predict or measure the success of the document. Published papers or company reports achieve success or failure in the minds of readers, and it is always difficult to predict whether a document will achieve the intended results.

The team should seek feedback from sources within the organization (especially if the document is an internal report) and/or outside of the organization when the document is published in the open literature. This will allow the team some measure of prediction of the success/failure of their work.

A word of caution must be added here. When documents appear in the open literature, authors often measure success by the number of times the published document has been cited by others in a specific time period. As appropriate as this may seem, many authors ignore the fact that the document was not cited because of its importance to the field but was cited because there were serious errors in, say, the experimental work that negated the data and any hypotheses drawn from the data.

7.5   RECOGNITION AND REWARD

In spite of teamwork being a combined effort of scientists or engineers, not only should the whole team be recognized for the work but also the individual role of each team member should be recognized and appreciated.

A productive team of scientists and/or engineers share common goals, a common vision, and have some level of interdependence that requires both verbal and physical interaction. Teams come into existence through shared attitudes about a particular writing project. They may come together for a number of different reasons, but their goals are the same—to achieve peak performance and experience success. The ends may differ but the means by which one gets there is the same—teamwork. Every member of the team is accountable when it comes to teamwork.

Teamwork has unique challenges and pressures that a system of team reward can successfully address, if correctly implemented (Klubnik, 1996; Armstrong, 2000; Parker et al., 2000).

To succeed at the task at hand, all involved need to combine their efforts. Everyone has his or her unique role, but each role must be recognized and appreciated.

Teamwork must be a high priority and given constant attention with every member understanding the importance of working smoothly together if they want to be successful. Furthermore, each team member must be dedicated to the whole team and be willing to act unselfishly. When challenges arise, the team needs to have the resources, accountability, and commitment to deal with them in a constructive and positive manner.

Recognition and reward are as serious to any scientist or engineer as they are to anyone in the workplace. It may be felt that the scientist or engineer has his or her reward by seeing a paper (with his or her name as a coauthor) in print or by being the coauthor of a well-accepted internal company report.

Recognition is about feeling special and it is often hard to feel special in a corporate program where everyone gets the same thing. Timing is important and the sooner an employee's performance is acknowledged, the more likely he or she is to repeat the desired performance.

Team rewards present a great opportunity to help foster team bonding. With the proper application of a reward and recognition program, team rewards can help build a high-performance team and foster cross-organizational cooperation. But team rewards are not to be treated carelessly as misapplication could lead to unhealthy competition, lack of cooperation, and ultimately severe financial consequences for the organization.

Team motivation and rewards programs should make up part of the overall employee recognition program. This may involve a mix of team and individual rewards with the mix balanced between awards that encourage both cooperative behavior and competitive behavior.

The desired outcome of recognition programs is to improve performance and improve employee retention.

In order to recognize successful teamwork, it is necessary to prioritize employee recognition and ensure a positive, productive, innovative organizational climate. There should be a means to provide employee recognition, to say *thank you,* and to encourage more of the actions and thinking you believe will make the team successful. People who feel appreciated are more positive about themselves and their ability to contribute. People with positive self-esteem are potentially the best employees and often, but not always, the best persons to work with others in a team. These beliefs about employee recognition are common among employers even though they are not generally executed. The result is that employee recognition is often a closely guarded secret in many organizations.

Time is an often-stated reason for the apparent lack of employee recognition and, admittedly, employee recognition does take time. Employers also start out with all of the best intentions when they seek to recognize employee performance. They often find their efforts turn into an opportunity for employee complaining, jealousy, and dissatisfaction which leave many employers hesitant to provide employee recognition.

Generally, employee recognition is scarce because of a combination of several factors. People do not know how to provide employee recognition effectively, so they have bad experiences when they do. They assume that *one size fits all* when employee recognition is provided.

Finally, employers think too narrowly about what people will find rewarding and recognizing.

Each organization should have *guidelines for effective employee recognition* but many organizations still use a scatter approach to this action. They talk a lot about employee recognition and hope that some efforts will stick and create the results they want. Or, they recognize it so infrequently that employee recognition becomes a downer for the many when the infrequent few are recognized.

Instead, organizations need to *create goals and action plans for employee recognition*, especially where teamwork is necessary. The organization needs to recognize the actions, behavior, approaches, and accomplishments that need to be fostered and reinforced as part of teamwork.

Employee recognition opportunities that emphasize and reinforce these sought-after qualities and behavior should be established. The team members need to see that each team member who makes the same or a similar contribution has an equal likelihood of receiving recognition for his or her efforts. For regularly provided employee recognition, organizations need to establish criteria for what makes a team member eligible for recognition. Above all, any team member who meets the criteria is then recognized. However, recognizing only the highest performer or one of the chosen few will spread dissatisfaction among all other team members, especially if the criteria for employee recognition are unclear or based on the team leader's opinion.

While there may be individual forms of recognition, it is important to recognize all people who contributed to a success equally. A message of thanks to the members of a successful writing project team, by the team leader, is always welcome, but if the leader misses the names of several people who contributed to the success of the project, this will only cause the excluded employees to feel slighted. Any subsequent thanks are meaningless, no matter how sincere.

Thus, the criteria of establishment of employee recognition awards need to be formally established in each organization that wishes to reward sterling performance. This is especially so when teamwork is required. Each member of the team must be mentioned and the procedure for selecting the award recipient must be standardized and transparent.

7.6 LEADERSHIP

The qualities of leadership—whether it is directing laboratory work or directing field work or writing—are essential for any team to be effective (Table 7.2). Some reference has been made to the team leader (above), but an expansion of these words will be helpful as the final section of this chapter.

Working as a member of a team can be rewarding or frustrating for the scientist or engineer. If there are poor communications from the team leader, team members may often feel confused and/or misunderstood. To create a successful team, effective communication methods are necessary for both team members and the team leader.

Effective communication skills enable the team leader to inspire and influence team members to reach the desired goal. A team leader with effective communication

---

TABLE 7.2

Characteristics of an Effective Team Leader

- Show respect for all team members.
- Ensure all team members understand the project goals.
- Hold regular team meetings to discuss the doubts and queries of team members.
- Keep every team member informed with the latest updates or issues.
- Define the role and responsibility of every team member to avoid confusion.
- In any meeting, allow everyone to speak and offer suggestions.
- Always offer sound and logical reasoning for opinions.
- Show appreciation when a team member performs well.
- Deal with tense situations calmly and professionally—avoid using harsh words.
- Avoid blaming others for any reason—attempt to determine the cause to ensure it is not repeated in the future.
- Through constant interaction with team members, develop a healthy team spirit.

skills allows the team leader to carry out his or her job with confidence, ease, and perfection. Effective communication skills can help a team leader in leaving a positive result with each interaction he or she has.

The team leader must be able to command energy, competence, and an ability to communicate as well as mediate among other team members. In addition, planning, initiating, organization, and controlling skills are required for team objectives to be realized. A team leader should serve the group and consider their first priority the collective interests of the whole. In other words, a team leader should not accept the appointment for his or her personal glorification.

Effective communication (including communication by the team leader) will produce results but effective teamwork certainly does not just happen automatically; it takes a great deal of hard work and compromise. There are a number of factors that must be in place to cohere together as a team and work seamlessly to produce an acceptable manuscript or report.

Effective leadership in the form of effective writing is one of the most important components of good teamwork. The leader should possess the skills to create and compile a manuscript as well as motivate and inspire his or her colleagues to make positive contributions to the work and be highly committed. A well-written manuscript will promote a high level of morale among the younger scientists and engineers and make them feel supported and valued.

Communication is a vital factor of all interpersonal interaction and especially that of a team working on a manuscript or report. It is also necessary for all coauthors to be able to articulate their feelings, express hypotheses, share ideas, and understand each other's viewpoints.

In this respect, conflict will undoubtedly arise no matter how well a team functions together. The best way to counter conflict in such cases is to allow team members to voice their concerns without fear of offending other members of the team. A hands-on approach by the team leader (who may also be the writer) that resolves such conflicts quickly is preferential. It is often advised that the team leader (and/or the writer) sit with the conflicting parties and help work out their differences without taking sides, trying to remain objective if possible.

In short, the team leader (writer) must set a good example in order to keep team members positive, committed, and motivated, and the team leader herself/himself needs to exhibit these qualities.

Above all, the team leader should have each contributor commit to completing the draft material within the expected timeframe. Since team members often have other responsibilities and busy schedules, the team leader must follow up with each author to ensure that the commitment is being met and find out whether any problems have occurred.

The key to creating a meaningful document through teamwork is for the team leader to make sure that all members of the team understand (1) the purpose the document is to serve and (2) the target audience who will read the document. With a thorough understanding of these two guiding principles, the team leader can make sure that the required level of detail is clear.

The process of developing a complex document is a leadership task that requires having clear expectations, commitments, and ongoing communication. Each step must be planned with time left for editing and revision. In all cases, the contents of the final document must be compared to the original purpose of the document and the requirements of the identified audience.

REFERENCES

Armstrong, M. 2000. *Rewarding Teams*. Chartered Institute of Personnel & Development, London, United Kingdom.

Bell, A.H. 1995. *Tools for Technical and Professional Communication*. NTC Publishing Group, Lincolnwood, Illinois.

Klubnik, J.P. 1996. *Rewarding and Recognizing Employees: Ideas for Individuals, Teams and Managers*. McGraw-Hill, New York.

Larson, C., and LaFasto, F.M.J. 1989. *Teamwork: What Must Go Right/What Can Go Wrong?* Sage Publications, Thousand Oaks, California.

Morgan, M., and Murray, M. 1991. Insight and Collaborative Writing. In *Collaborative Writing in Industry: Investigations in Theory and Practice*. M.M. Lay and W.M. Karis (Editors). Baywood Publishing Company, Amityville Publishing Company, New York. Pp. 64–81.

Parker, G.M., Zielinski, D., and McAdams, J. 2000. *Rewarding Teams: Lessons from the Trenches*. Jossey-Bass, John Wiley & Sons, Hoboken, New Jersey.

Peter, L.J., and Hull R. 1969. *The Peter Principle: Why Things Always Go Wrong*. William Morrow and Company, New York.

Spack, R. 1988. Initiating ESL students into the academic discourse community: How far should we go? *TESOL Quarterly*, 22, 29–53.

Speight, J.G., and Foote, R. 2011. *Ethics in Science and Engineering*. Scrivener Publishing, Salem, Massachusetts.

8 Publishing

8.1 INTRODUCTION

Publication plays a critical role in the advancement of science and engineering by communicating knowledge from the researcher(s) to the larger scientific community as well as advancing the support of various companies for research (Jameson, 1995; Davis, 1997).

In fact, there are those purists who argue that science and engineering research does not exist until the work is published, at which time the publication becomes a public commodity. The exchange of information through publication is an essential part of doing science and engineering, a public good, and, for some, a moral imperative. It is important, then, that scientific societies, as major publishers of science and engineering, take initiatives to preserve the integrity of the process that certifies and communicates research (Beckett, 2003).

Publication of research work is essential in order to advance science and to engineering papers. It is also essential for people pursuing a scientific career. The recognition of scientists and engineers as researchers depends on their publications and contributions to scientific progress. Many scientists and engineers live in a *publish or perish* world and must produce a certain number of publications within a specified time frame. On the other hand, the commercial world, while not specifying published works in the open literature, expects scientific and engineering researchers to publish internal company reports that advertise the researchers' work to the whole company. It is from such reports that intellectual property (patentable material) arises thereby giving the company an edge on market competitors.

For the typical researcher who is either (1) an academic scientist or engineer for whom publication is life or (2) a company scientist or engineer who may publish on occasion, publication of papers in peer-reviewed journals is the predominant form of publication for scientists and engineers, and usually such journals have the highest readership. However, journals vary enormously in their prestige and importance, and the value of a published article depends on the journal.

Peer review is a general term that is used to describe a process of self-regulation by scientists and engineers (as well as for many other professions) as a means of evaluation of a paper before publication and involves review by qualified individuals in the relevant scientific or engineering field. Peer review methods are employed to maintain standards, improve performance, verify whether the work satisfies the specifications for review, identify any deviations from the standards, and provide suggestions for improvements.

Thus, researchers should learn not only how to write a scientific or engineering paper, but also how to get it published. Scientific and engineering journals have technical requirements, and authors should make themselves familiar with these requirements (Chapter 2). Scientists and engineers deserve to have the credit for their work, but this applies only if it is their work and they have made an intellectual contribution to the project. Ethical standards (Chapter 5) apply to scientific publication and should be observed by authors, and ensured by journal editors.

Typically, when publishing the results of their work, scientific and engineering professionals who are biased toward theory tend to produce data that are often abstract and the intellectual contribution is expressed in the form of theories with proof. As a result, publication on the proceedings of a conference may be the only outlet for their efforts after which publication in a *reputable* journal may be possible but only with considerable efforts or, for various reasons, may not be possible at all.

For the nonacademic scientist and engineer, there is the medium of publication of the material as a *company report*. This can be a worthwhile method for circulating one's work throughout the company. However, the importance of the work to the young scientist and engineer can, again, be diminished and the names of a supervisor and any other persons higher up the food chain are included as coauthors.

Publication of data in the proceedings from a conference often results in a shorter time to print. This follows from the opportunity to describe completed or partly completed work before peer scientists and/or engineers and to receive a more complete review than the type of review that is typical for a journal. At a conference, the audience asks general and specific questions to the presenter that often provides recommendations for further work or a new line of investigation. Overall, this will help the presenter to finalize the document for publication in the proceedings (where the proceedings are published post-conference).

On the other hand, one has to wonder if journal reviewers really pay attention to the salient points of the potential publication or do they merely look for errors in style and grammar. An answer that several readers may relate to is *all of the above*. However, in many academic reviews, statements are made that publication in the proceedings of a prestigious conference is inferior to publication in a prestigious journal without realizing or being willing to admit that in relation to data presentation and publication, many conferences are superior to an established journal.

However, publication of research data is not an open form of recognition for all scientists or engineers. Scientists and engineers employed in industry may be prevented from publishing their work because of (1) a company policy related to proprietary material—a justified reason—or (2) an arbitrary decision by a supervisor or a member of the company review committee—not a justified reason.

On the other hand, in academia, the young professional enters a department at the assistant professor grade. At this level, the assistant professor has little choice in terms of choosing teaching assignments and has administrative work thrust upon his or her shoulders while the older tenured members of staff often refuse such work—using any lame excuse that comes to mind—without fear of reprisal. This is surely a breach of ethics, which may involve untruths or merely laziness.

In addition, the young assistant professor also has to acquire research funding and may even have to pass his or her reports/papers through a review committee prior to publication. This review committee will be made up of senior members of staff who, for many reasons that are often difficult to follow, can give the young professor a glowing performance report or a report that is somewhat less than glowing. It is at this time, if the latter is the case, that the young assistant professor can feel that he or she is suffering rejection by one's colleagues.

The educated young professional scientist and engineer wonders if he or she is merely a pair of hands (for an overbearing supervisor or an overbearing department head or jealous colleagues) and not supposed to be given credit for the ability to think and solve a problem. Performance suffers and, with repeated negativism toward publication, the young professional starts to lose interest in the organization.

Lack of recognition for hard and intelligent work is a killer and getting the best out of any such scientists and engineers becomes an impossible dream.

There is an extremely important role for the scientific societies in developing authorship policies for their members. The societies must also make sure that their members know of the existence of their policies and how to interpret them. Regular continuing educational efforts are imperative. There is also the possibility that scientific societies could work together to establish a uniform policy that would hold across disciplines. This would be advantageous to those engaged in interdisciplinary research collaboration.

While the general definition of scientific misconduct includes fabrication, falsification, and plagiarism, the scientific community is charged with considering standards for other practices. In publication practices, that encompasses such matters as authorship credit, duplicate publication, accurate representations of the data presented, and peer review.

Generally, the following criteria need to be observed when compiling data for publication: (1) all persons designated as authors should qualify for authorship, and (2) each author should have participated sufficiently in the work to take public responsibility for the content. Furthermore, authorship credit should be based only on substantial contributions to (1) either the conception and design or the analysis and interpretation of data; (2) drafting the article or revising it critically for important intellectual content; and (3) final approval of the version to be published. Conditions 1, 2, and 3 must all be met. Other contributors should be listed in an appendix or footnote. Editors may ask authors to describe their contribution(s).

However, publishing is undergoing redefinition as electronic publications and there are both opportunities and pitfalls associated with electronic publishing. The global reach of electronic publishing mean that new, expanded audiences can be reached. In addition, digital technology may make it easier to misrepresent data or alter graphic representations. Scientific and engineering societies could make a valuable contribution by encouraging cross-disciplinary discussion of these matters among researchers and those involved in publishing.

Guidelines for responsible conduct in the electronic communication and electronic publication of scientific research must be developed and implemented, and the societies can play a pivotal role in their promulgation and implementation.

Generally, publications are variable but, in general, constitute: (1) a report and a record of the activity of researchers, (2) the references research builds upon, (3) the data governments, organizations, or society can refer to when facing issues that have a major scientific component, and (4) a way of evaluating the scientific and engineering activities of individual researchers. The quality of publications, in terms of scientific integrity, is therefore essential for research to be conducted in an efficient and responsible way and for a transparent communication between researchers and society.

Science and engineering disciplines are flourishing because of communication, a much broader concept than publishing. Hopefully, before the article will be published, researchers will have had extensive discussions with their peers to share their views, ideas, and opinions in order to check the validity of their claims. Unfortunately, in recent times the advancement of knowledge has not appeared to be a top priority for many scientists and engineers. Fame and fortune have become the focus of the researchers and publication of data that have not been confirmed or data that have been made up have seen the light of day in scientific and engineering journals.

Many opportunities and concerns are in play in scholarly publication and communication. These result from capabilities afforded by new technologies, pressures associated with the publish-or-perish message that is forced on many scientists and engineers in academia, or the invent-or-perish message that is forced on many industrial scientists and engineers.

While the unethical behavior of scientists and engineers cannot be blamed on the publish-or-perish message or on the invent-or-perish message, the pressure placed on the shoulders of many individuals by either of these messages may be a contributing factor. Not that anyone found guilty of unethical behavior should be excused because of such a message, but it may be time to change the message—if that is at all possible.

The young assistant professor who is excused from being reprimanded or punished for unethical behavior because he or she is a young professor seeking funding for a research project is also not a valid excuse for lack of disciplinary action. In fact, one might ask if those exalted academics promoting and accepting such an excuse are not also guilty of unethical behavior because they have condoned the professor's actions.

In fact, the lack of willingness of the (academic or industrial) faculty to change is a key barrier to reducing and perhaps eliminating unethical behavior in science and engineering.

While there are claims that gross scientific misconduct are assumed to be rare, subtler forms of unethical behavior are becoming more common (Ritter, 2001; Speight and Foote, 2011). Misappropriated credit in publications, for example, can lead to some of the most contentious conflicts in the academic world. Currently, in academia, publication of research data has become more competitive because universities and organizations are more focused on intellectual property and rights of ownership. In addition, research that is sponsored by commercial entities is usually controlled, and it is the commercial organization that determines whether and how results are published; it is no longer an academic issue.

Guidelines for responsible conduct in the communication and publication of scientific research must be developed and implemented, and the societies can play a pivotal role in their promulgation and implementation.

The need to disseminate scientific and engineering knowledge and expertise for the betterment of the discipline led to the concept of publishing the results of work in scientific and engineering journals.

Scientific and engineering publications that resulted from such a concept provided mediums for the spread of knowledge as well as additional benefits such as reputation among peers and even monetary benefits. In fact, publishing the results of scientific or engineering work in the relevant journals regularly is often a prerequisite for appointment or promotion in several institutions (especially universities). Furthermore, scientific or engineering publications are a compulsory requirement for appointments to senior positions. Thus, with publications becoming synonymous with job survival, many scientists and engineers have started publishing aggressively in the recent years.

8.2   TYPES OF JOURNALS

Once the decision has been made to incorporate the results of scientific or engineering work into a paper the journal that seems most appropriate to the work must be selected. In fact, the need to disseminate scientific and engineering knowledge and expertise for the betterment of the discipline led to the concept of publishing the results of work in scientific and engineering journals.

Scientific and engineering publications that resulted from such a concept provided mediums for the spread of knowledge as well as additional benefits such as reputation among peers and even monetary benefits. In fact, publishing the results of scientific or engineering work in the relevant journals regularly is often a prerequisite for appointment or promotion in several institutions (especially universities). Furthermore, scientific or engineering publications are a compulsory requirement for appointments to senior positions. Thus, with publications becoming synonymous with job survival, many scientists and engineers have started publishing aggressively in recent years.

In the present context, the scientific and engineering literature comprises scientific and engineering journals that report original empirical and theoretical work in scientific and engineering disciplines. University researchers favor publication in such journals while their industrial counterparts may have to focus on patents.

However, before submitting a manuscript for publication in a journal, it is important to consider the functions of different journals. All journals strive to publish the best science or engineering but the areas that they cover differ. Some journals are broad in their coverage of scientific and engineering disciplines while other journals deal with more specific areas of science or engineering. Still other journals publish work from the members of the parent scientific society or engineering society and, in such cases, publication may be a *members only* policy.

Currently, peer-reviewed journal articles remain the predominant publication type, and have the highest prestige. However, journals vary enormously in their prestige and importance, and the value of a published article depends on the journal. The status of conference proceedings depends on the discipline; they are typically more important in the applied sciences and engineering, especially for industrial scientists and engineers.

In many scientific and engineering disciplines, advancement depends upon publishing in so-called high-impact journals, most of which are English-language journals. Scientists and engineers with poor English writing skills are at a disadvantage when trying to publish in these journals, regardless of the quality of the scientific study itself. Yet many international universities require publication in these high-impact journals by both their students and faculty. One way that some international authors are beginning to overcome this problem is by working with technical copy editors who are native speakers of English and specialize in editing texts written by authors whose native language is not English to improve the written quality to a level that high-impact journals will accept.

For the publishing purist, journals can be conveniently (or must be necessarily) divided into four general classes: (1) primary journals, (2) secondary journals, (3) society journals, and (4) specialty journals.

The *primary journals* attempt to publish articles that make conclusions that shift the current theory the current line of thinking or modify the current paradigm in a field. Articles in these journals are typically highly focused and may not provide in-depth coverage of the area. These journals, like the authors contributing the articles, are often concerned with being the *first* to publish information that may be of interest to the reading audience.

The *secondary journals* typically represent specific fields or general areas of science and engineering. Over the past three decades, some of the primary journals have created spinoff journals that specialize in a particular area. These journals also prefer manuscripts dealing with cutting-edge technology that include complete and in-depth description.

Society journals publish the bulk of the work coming out of most laboratories and are the workhorses of the scientific publishing industry. These journals may have more citations per year than the primary journals and secondary journals and the work published in them has a very high quality.

Specialty journals publish work in a restricted area. Although this work may be of high quality, too, the readership may be limited to only those in the field.

No matter how they are categorized, all journals value (1) novelty and unexpected findings and/or (2) careful, extensive analyses of critical scientific or engineering processes.

The would-be author is well advised to survey the various journals related to his or her field of endeavor and locate the journal that is the *best fit* for the work. Invaluable advice for the most suitable journal can be available from colleagues and other scientists or engineers in the field who have experience as authors, reviewers, and journal editors.

A major fact that is being forgotten about the quality of a journal is that it depends on the quality of manuscripts submitted by the authors, qualified referees, standardized peer-reviewing, and suitable and demanding readers, more than the editor and his or her team.

It may be tempting for a young scientist or engineer to send his or her manuscript to a so-called primary journal even when the results are not of the highest novelty or broadest interest. This will only lead to frustration when the manuscript is rejected.

There is a *school of thought* that it does not really matter what journal a scientist or engineer's work is published in, because if the science or engineering is good enough other scientists and engineers will find it, read it, and cite it. This statement is, to a large extent, true today, because of the ease with which the literature can be searched and articles can be downloaded, which saves scientists and engineers the effort of getting lost in library stacks. However, the author should ensure that the work is published at least at the level of scientific or engineering society journals, especially if the work is closer to the cutting edge.

The manuscript should be sent to the most appropriate journal first instead of waiting until it is rejected by an inappropriate journal.

The impact of an article or journal can be measured directly by the number of times the average article is cited in other articles. This number—the *impact factor*—is a measure often used by chairpersons (or tenure dossier reviewers) to measure the prowess of a faculty member coming up for promotion. The (sometimes erroneous) assumption here is that articles published in journals with high impact factors count more than do articles published in journals with lower impact factors.

The *impact factor* is the number of citations a journal receives in a given year for articles it published over the previous two-year period, divided by the number of articles it published in that period. This provides comparisons between journals, and it (the impact factor) serves as an indicator of the number of times a particular article has been cited. Thus, the higher the number, the higher the journal's ranking.

In some cases, the impact factor may also serve as a gauge of quality, a development that many view as problematic—as long as the impact factor is not a measure invented by one publisher for application to journals published by that particular company. For example, what does the impact factor really mean if papers from journals of Publisher A have higher impact factors when judged using the system that originated from Publisher A?

8.3 JOURNAL EDITOR

Descriptions of duties and responsibilities of an editor of a scientific or engineering journal are not readily available—not because they are hidden but because they are not usually defined in writing.

Generally, the editor of a peer-reviewed journal is responsible for deciding which of the articles submitted to the journal should be published, often working in conjunction with the relevant society (for society-owned or sponsored journals). The validation of the work in question and its importance to researchers and readers must always drive such decisions. The editor may be guided by the policies of the Editorial Board of the journal or constrained by such legal requirements as shall then be in force regarding libel, copyright infringement, and plagiarism.

More specifically, the journal editor (or editor in chief, as the title may indicate) has final decision-making authority on, and is responsible for, the appropriate delegation of Editorial Board responsibilities related to the scientific, engineering, and other editorial content of the journal, including solicitation and acceptance or rejection of manuscripts; selection of editorial boards and reviewers; and the approach to correspondence with authors, reviewers, and readers.

In order to accept these duties and responsibilities, it is preferable that the editor is skilled in the areas of scholarship covered by the journal. In short, the editor of a scientific or engineering journal has the added responsibility to check the technical soundness and technical quality of the content. For this, the editor is required to have the technical skills and up-to-date knowledge of the area of scholarship covered by the subject matter of the journal.

Thus, the assumption (often correct, but not always) is that the editor chosen for a journal is the most appropriate scientist or engineer because the editor is the *gatekeeper* or *watchdog* for the journal and anything published in the journal must pass across the editor's desk and *must* be reviewed by the editor. Authors who submit manuscripts to the journal for possible publication are often swayed by the qualification and knowledge of the editor—the reputation of the editor is a direct influence on the reputation of the journal.

The editor is, in fact, the *quality control officer* for the journal where a check is made on the content (authenticity and relevancy to the topic), language (grammar and content flow), and aesthetics (photos, images, sound, audio, and video) of the articles or documents appearing on the specified medium. The job of the Editor also involves relationship building and communication with the author. He or she is also required to use his or her creative and human resource skills to maintain a cordial relationship with authors whose articles are rejected.

The success of a journal will depend on the performance of the editor and it is the job of the editor to shepherd the journal through lean times and through good times (McHugh, 1998). For those who seek to build a résumé by including journal editorship as a glowing one-liner, it is recommended that they seek early retirement.

The other major responsibility of the editor is the administration of the peer review process, which assists him/her in making editorial decisions, and through the editorial communications with the author, may also assist the author in improving the paper. All scientists and engineers who wish to contribute to publications should be willing (perhaps even have an obligation) to act as peer reviewers.

Furthermore, the editor should also know the reviewers sufficiently well to know that they are qualified, impartial, and fair, and the confidentiality of the review process must be preserved.

Protecting intellectual property is a primary responsibility of the editor. The editor should know the reviewers well enough so that any thoughts of plagiarizing manuscripts by reviewers should not be an issue. Reviewers must not use ideas from or show another person the manuscript they have been asked to review without the explicit permission, via the journal editor, of the author of the manuscript. Advice regarding specific, limited aspects of the manuscript may be sought from colleagues with specific expertise, provided the author's identity and intellectual property remain secure.

The editor must accept it as a hard-and-fast rule (a rule without exception) that unpublished materials disclosed in a submitted manuscript must not be used in a reviewer's own research without the express written consent of the author. Privileged information or ideas obtained through peer review must be kept confidential and not used for personal advantage. Reviewers should not consider manuscripts in which they have conflicts of interest resulting from competitive, collaborative, or other relationships or connections with any of the authors, companies, or institutions connected to the submitted unpublished papers.

The editor must also maintain reviewing schedules by ensuring that reviewers meet agreed-upon reviewing deadlines. And, in order to maintain good reviewers the editor should have the means to evaluate performance of editorial review board members and coach when appropriate.

While some editors consider themselves to be the all-powerful authority for publication in the journal, the editor should be willing to step down from this lofty perch and assist authors in developing articles to the fullest potential. To do this, the editor must maintain all communications with all authors and reviewers in a courteous and diplomatic manner. This also involves correspondence sent to authors in relation to checking every manuscript for completeness, references, tables, figures suitable for reproduction, legends, abstracts, permission to use copyrighted material, and mailing address for proof.

Copyright should be respected and copyright protection begins at the time a creative work is recorded in some tangible form. In many scientific and engineering disciplines, there is seldom a financial compensation for copyright, as in other fields, but there is certainly the need for recognition. No figure or table from previously published work should be included without written permission from the publisher or without acknowledgment of the source of the work from which the figure or table is taken.

Above all, and certainly in the context of the present text, the editor must not be involved in positions where conflicts of interest can arise, so that all decisions are beyond reproach.

A conflict of interest may exist when a manuscript under review puts forth a position contrary to the reviewer's published work or when a manuscript author or reviewer has a substantial direct or indirect financial interest in the subject matter of the manuscript. A conflict of interest may also exist when a reviewer knows the author of a manuscript. The editor should ensure that such conflicts do not occur and that he or she is also beyond reproach.

The editor should also ensure that proper acknowledgment of the work of others is given. Authors should cite publications that have been influential in determining the nature of the reported work. Information obtained privately, as in conversation, correspondence, or discussion with third parties, must not be used or reported without explicit, written permission from the source. Information obtained in the course of confidential services, such as refereeing manuscripts or grant applications, must not be used without the explicit written permission of the author of the work involved in these services.

The editor should recognize that a conflict of interest does not exist when an author disagrees with a reviewer's assessment that a problem is unimportant or disagrees with an editorial outcome.

The journal editors should also take all reasonable steps to ensure the accuracy of the material that will be published in the journal. Indeed, whenever a significant inaccuracy, misleading statement, or distorted report is recognized in a published paper, it must be corrected promptly. If articles prove to be fraudulent or contain major errors, they should be retracted (with the word *retraction* used in the title of the retraction to ensure that it is picked up by indexing systems). On the other hand, meaningful critical responses to published material should be published unless editors have convincing reasons why the responses cannot be published.

In short, the editor should take reasonably responsive measures when ethical complaints have been presented concerning a submitted manuscript or published paper, in conjunction with the publisher. Such measures will generally include contacting the author of the manuscript or paper and giving due consideration of the respective complaint or claims made. Actions taken by the editor may also include further communications to the relevant institutions and research bodies and, if the complaint is upheld, the publication of a correction, retraction, expression of concern, or other note is required. Every reported act of unethical publishing behavior must be investigated, even if it is discovered years after publication.

8.4 PEER REVIEW PROCESS

Peer review is the means by which scientific and technical manuscripts submitted for publication in journals undergo *quality control* in the form of a check on technical quality, the lack of flaws in the data, and the validity of the conclusions drawn from the data. There is also a criterion that is often forgotten and that relates to the qualifications of the peer reviewer and whether or not the system works (Lock, 1994). For example, if the reviewers of a submitted manuscript on tar sand technology are scientists and/or engineers with no experience in that area of technology, the purpose of the peer review system is defeated whether or not the paper is rejected or accepted (Grant and Laird, 1999).

The job of the peer reviewer is to determine, through an honest and objective review, if the work is suitable for publication in the journal to which it is submitted (Hoppin, 2002). Different journals have different criteria that the work has to meet. Reviewers must justify their opinions on acceptance, revision, or rejection of each manuscript. Reviewers justify their recommendations to the author through their critique of the work.

Any selected reviewer who feels unqualified to review the research reported in a manuscript or knows that its prompt review will be impossible should notify the editor and excuse himself or herself from the review process.

In addition, the journal peer review process has three purposes. The first is quality control, to eliminate major errors in papers and unsuitable papers. Second, the review process should ensure fair treatment of all authors (not just for cronies of the editor and especially not for paper authored or coauthored by the editor). Third, the review process encourages the publication of papers that contain new and useful findings.

A cornerstone of science and engineering is that the findings must be reproducible and well documented. Many instances of scientific misconduct have been exposed when other scientists cannot reproduce reported data (Speight and Foote, 2011). When other scientists or engineers are unable to replicate the work, it is often discredited. To keep this type of conduct in check, scientific and engineering articles usually include detailed descriptions of experimental protocols that enable others to reproduce the experiments.

The most basic distinction between journal review processes is whether manuscripts are triaged. Triaging processes are used by editors who attempt to prescreen manuscripts so that they send out for review only those that have a chance of being selected for publication.

Several different systems of peer review are employed and each journal has its own system in place. Most systems use two or more peer reviewers to assess the suitability of an article for publication.

For the author, the three most important aspects of the peer review system are (1) the quality of the review, (2) the length of time it takes for the review process to be completed, and (3) once accepted, the length of time before a manuscript goes into print. Many scientists and engineers place less emphasis on the last point because, once accepted, the manuscript is given a DOI number (*Digital Object Identifier*), which is a unique identification code that allows an article to be easily tracked by many different archival and research programs on the Internet. Every article that is accepted for publication in a recognized journal should be assigned a DOI number—it is the responsibility of the copy editor or production editor to make sure that this number is formed correctly.

To the scientist or engineer, the peer review process is the benchmark of quality and results from external and independent valuation. It also functions as an effective means of winnowing the papers that a researcher needs to examine in the course of his or her research (Harley et al., 2007).

The peer-review process is more complicated for compound disciplines that cross over between science and engineering because many such fields are relatively new and therefore result in small, specialized communities of scholarship.

Scientists and engineers in these interdisciplinary fields often prefer to publish within a single traditional discipline because the most highly respected and recognizable outlets reside there; however, divergent expectations (ranging from quantity to methodology to writing style) and standards (especially with regard to quality) among fields often make it difficult for reviewers in standard fields to judge submissions from compound disciplines. Interdisciplinary publications may address this concern more readily as they become more prestigious.

Reviewers are chosen by the editor on the basis of their expertise in the field, often utilizing extensive databases assembled by the journal and the editor's knowledge of the area. Some scientists are better reviewers than others—they are more critical and thorough, a fact that quickly becomes known to editors. The identities of the reviewers are generally not revealed to manuscript authors. This later rule is intended to free reviewers from any social pressures, allowing them to consider only the quality of the science before them.

Reviewers consider (1) the validity of the approach, (2) the originality of the finding, (3) the significance of the findings, (4) the interest and timeliness of the work to the scientific community and the engineering community, and (5) the clarity of the writing.

The review process can take anywhere from a few days to several weeks. Once the review is complete, each reviewer provides feedback on the manuscript to the journal editor, who relies on peer-review feedback to guide publication decisions.

After reports of the reviewers are received, the editor makes a decision about publication, taking into account all of the feedback he or she has received. The editorial goals of the journal—sometimes journal editors decide that certain areas are of particular upcoming or lessening interest—factor into the decision, as does knowledge about the reviewers themselves and the background behind their opinions.

Authors use reviewers' comments to refine the text of their manuscript and the experiments within. Journal editors must occasionally resolve issues related to conflict of interest among reviewers.

The peer-review system allows readers to feel confident that the information found in scientific journals is credible. Despite the best efforts of reviewers, cases of scientific misconduct do occur and incorrect or unsubstantiated data does get published (Speight and Foote, 2011).

Finally, reviewers are expected to keep the information in a manuscript confidential until it is published, but it is rare that the work comes as a complete surprise to the entire scientific community. This is because peer review is integrated into almost every step of science. Many research scientists and engineers request public funding for their experiments. Funding decisions are made by a committee of other scientists or engineers who debate each proposal's likelihood of success, the validity of its approach, and the importance of the question being asked. Once funded, the experiments can begin, and preliminary data are often revealed at scientific meetings. This allows the findings to be debated and defended with colleagues prior to publication. Once the experiments are completed, a manuscript is written and circulated to all those who contributed to the work.

8.5   ASSESSING REVIEWER'S COMMENTS

The peer reviewer(s) expect(s) that the author will, at the very least, address his or her comments. This is very important, since most revisions are rereviewed by the same peer reviewers who will be checking to determine if the author(s) acknowledged and considered the comments.

Response letters should state that the author thanks the reviewers for their time and effort and their contributions to the work and that by addressing the comments of the reviewers the work was improved.

If the author decided to do nothing, he or she (the author) needs to provide a point-by-point response to each reviewer's concerns. If the author's response is supported by the published works, supplying references (with necessary quotations) will strengthen his or her point.

An author can rebut a rejection decision by writing a letter to the editor to explain why he or she believes that the reviewers came to the wrong decision. The author should not blame the reviewers for not doing their job. If the rejection was based on a misinterpretation of the results by the reviewer(s), there is a chance that the editor will request that the manuscript be reviewed again by the same reviewers or by a different panel of reviewers.

In places where the author and the reviewer(s) agree, the author should note in the manuscript where revisions have been in accordance with the reviewer's comments/questions. This will help the editor and the reviewer (if the manuscript is sent out for rereview) to locate any changes and determine if the changes have addressed the issues.

Be advised that point-by-point letters are often very long and the author should be as succinct as possible while also being clear—derogatory remarks about the review should be avoided. However, the author can send a letter to the editor to explain why he or she (the author) believes the reviewers came to the wrong decision.

Yet, the author must recognize that there is a good chance that his or her rebuttal will be rebutted.

The most appropriate action is to make the changes suggested by the reviewers and submit the manuscript to another journal.

Reviewers may make comments about the way a figure or table is assembled or presented. In fact, if a reviewer comments that data are not presented clearly, the author should make the necessary changes/corrections. However, to avoid such actions, it is wise for the author to ask colleagues for their suggestions before the paper is submitted to the journal.

In addition, receiving a comment that the work needs to be edited by someone who speaks and writes English as his or her primary language is one that should be taken seriously.

8.5.1   ASSESSING COMMENTS ON A REJECTED MANUSCRIPT

If the manuscript was rejected, the first question the author(s) must ask is "Why?" and relate this question to the categories listed above.

Typically, rejections based on novelty or significance and relevance to a field indicate that the paper was submitted to the wrong journal. To correct this, the author must reassess the work and choose a more appropriate journal.

Reviewers typically comment on the experimental design and the quality of the data. While every scientist or engineer has a different way of performing an experiment, in the end the data and method of data collection must support the conclusions that are drawn from the data. Two fundamental relevant questions that the author must ask himself/herself are as follows:

- Do the data support the interpretation?
- Is the interpretation the only interpretation that the data support?

If the reviewer comments on any of these issues, the author needs to address them either by (1) providing justification for the work, (2) performing additional experiments, (3) providing more information, (4) providing more detailed discussion of the data, or (5) all of the aforementioned before resubmitting the manuscript.

As hard as it is to receive a rejection letter, the key to success in science is to receive criticism as openly as possible and without bitterness, and to respond by incorporating or debating the critique in the revised manuscript. The key for the author is to know how to move forward, to understand the appropriate roles of the reviewer and the reviewed manuscript, and to determine how to respond to critiques so that the paper will be accepted.

The open door is that the author can revise the work and resubmit it to another journal.

8.5.2   ASSESSING COMMENTS REQUESTING A MAJOR REVISION

Many journals provide authors with a second or, sometimes, a third chance to get their work accepted—typically when the work is sound and interesting to the reviewer but is incomplete in either experimentation or interpretation.

The letter from the editor may state that the work is not acceptable in its current form or give the author the option to submit a revised manuscript after more work has been done. The editor will assume that the author will follow the reviewers' advice whenever possible.

Responding to experimental design and/or issues related to data takes time. To minimize the amount of time and effort, the author must consider the most important points that the reviewer wants addressed experimentally. It is, however, necessary for the author to respond to the editor and advise him/her of the decision to proceed or withdraw the manuscript.

This sends a message to the editor and to the reviewer that the author is willing to follow their advice. Such a response might be looked upon very favorably.

8.5.3 Assessing Comments Requesting a Minor Revision

Often, an author might receive a letter from the journal editor indicating that the manuscript will be accepted, subject to satisfactory responses to the reviewers' comments.

The comments can usually be addressed by (1) acknowledging the reviewers' interpretation in the manuscript, or (2) adjusting a sentence here and there.

If a reviewer suggests that an author change some of the wording to make it more presentable, the author should agree to this course of action. In the end, the copy editor or production editor may change the wording to something completely different.

8.5.4 Discovery of Errors

When an author discovers a significant error or inaccuracy in his or her own published work, it is the author's obligation to promptly notify the journal editor or publisher and cooperate with the editor to retract or correct the paper. If the editor or the publisher learns from a third party that a published work contains a significant error, it is the obligation of the author to promptly retract or correct the paper or provide evidence to the editor of the correctness of the original paper.

8.6 FRAUD

Although errors do occur in scientist and engineering published papers after which the author notifies the editor to make corrections, misconduct (fraudulent presentation of data or fraudulent use of the data of other workers) is a much more serious issue.

Misconduct in scientific and engineering research can range from errors of judgment (such as, but not limited to, biased use of analytical data as well as misuse of statistical analysis) to falsification (omitting data—leaving out data points that do not fit the researcher's hypothesis—data manipulation, data exclusion, suppression of inconvenient facts) to blatant fraud, usually categorized as fabrication, falsification and plagiarism (Farthing, 1998).

Plagiarism is a major ethical offence and falls under the category of fraud. For a scientist or engineer to use and claim the words or ideas of another person as his or her own, without acknowledging the contribution of the other person, is unacceptable.

The cultures of science and engineering are based on accuracy and trust (Speight and Foote, 2011). When a researcher presents his or her data to the scientific and engineering public, the data are taken at face value and are rarely questioned—unless there are glaring errors that jump from the page and hit the reader in the face.

Other scientists and engineers may interpret data differently, question the study design, or disagree with the statistical analysis, but this is merely readers voicing their own thoughts and opinions on the data. It does not mean that the data has been manipulated or changed to fit the writer's hypothesis. Disagreement (or agreement) is the essence of scientific and engineering thought.

However, if the data cannot be trusted, for whatever reason, the concepts of science and engineering are compromised, even destroyed. When such cases occur and are published in the media, public confidence in science is destroyed. Scientific and engineering fraud also indicates that public or company funds may have been abused. Cases of scientific and engineering fraud are causes for embarrassment and frustration to the vast majority of honest scientific researchers (Speight and Foote, 2011).

Finally, another form of misconduct (although not generally recognized as such) involves scientific and engineering researchers who deliberately leak news of their findings to the public or the media before they have communicated the discovery to their respective peers in scientific press or at meetings. In fact, the preliminary release of scientific or engineering information described in a paper that has been accepted but not yet published violates the policies of many journals. In exceptional cases, and typically by arrangement with the journal editor, preliminary release of data may be acceptable.

REFERENCES

Beckett, R. 2003. *Communication Ethics: Principle and Practice. Journal of Communication Management*, Henry Stewart Conferences and Publications, London, 8(1): 43–44.

Davis, M. 1997. *Scientific Papers and Presentations*. Academic Press, New York.

Farthing M.J.G. 1998. Ethics of Publication. In *How to Write a Paper*. 2nd edition. George M. Hall (Editor). BMJ Books, London, United Kingdom. Pp. 148–158.

Grant, J.M., and Laird A. 1999. How a paper is reviewed and why it might be turned down. In *Introduction to Research Methodology for Specialists and Trainees*. P.M.S. O'Brien and F.B. Broughton-Pipkin (Editors). Royal College of Obstetricians and Gynecologists Press, London, United Kingdom. Pp. 231–240.

Harley, D., Earl-Novell, S., Arter, J., Lawrence, S., and King, C.J. 2007. The Influence of Academic Values on Scholarly Publication and Communication Practices. *Journal of Electronic Publishing*, 10(2).

Hoppin, F.G. Jr. 2002. How I review an Original Scientific Article. *American Journal of Respiratory and Critical Care Medicine*, 166: 1019–1023.

Jameson, M. 1995. The Benefits of Publishing Technical Papers. *Fire Technology*, 31(4): 372–374.

Lock. S. 1994. Does Editorial Peer Review Work? *Ann. Intern Med.*, 121: 60–61.

McHugh, J.B. 1998. *Responsibilities of a Journal Editor*. McHugh Publishing Reprint 22. http://www.johnbmchugh.com/pdfs/P-2%2022%20Responsibilities%20of%20a%20Journal%20Editor.pdf

Ritter, S.K. 2001. Publication Ethics: Rights and Wrongs. *Chemical & Engineering News*, November 12: 24–31.

Speight, J.G., and Foote, R. 2011. *Ethics in Science and Engineering*. Scrivener Publishing, Salem, Massachusetts.

9 Technical Presentations

9.1 INTRODUCTION

There will be many times in the career of a scientist or engineer when he or she will be called upon to present a paper at a scientific or engineering conference or even to company management. While excellence in research is often the main criterion for success as a scientist or engineer, reputation as a competent speaker can take a career to higher levels in leaps and bounds. A competent speaker will more likely be invited to give invited talks at important conferences or within the company system (Parberry, 1989).

Furthermore, quality of the presentation can be a deciding factor in the selection or rejection of a controversial conference paper or in the promotion of the speaker.

Good presentation skills are a prerequisite of a successful scientist or engineer (Parberry, 1989; Mandel, 1994; Li, 1999; Alley, 2003; Walker 2010). Good ideas will not be recognized unless they are effectively conveyed to others. The presentation should be rehearsed because this will help determine how many slides should be included in the presentation. A good rule of thumb is one slide per minute of presentation, although that depends on the individual speaker and the contents of the slide.

It should be no surprise that before a scientist or engineer starts to prepare a presentation (sometimes called *a talk*) he or she must know the goal of the presentation (i.e., what points must be made and understood by the audience) and know his or her audience. The presenter must customize his or her presentation to the audience and even if the presentation was previously created for another venue, a new presentation may be necessary.

The goal of a presentation is similar to the goal of a scientific or engineering paper or report. In either case, the audience needs to be convinced that (1) the research problem is worthwhile and relevant and a solution would be useful, (2) the problem has not been solved by others, (3) there are no other ways to achieve equally good results, and (4) the presentation will provide a solution to the problem. If any of these parts are missing, the presentation is less likely to be a success. The presenter must be sure to provide motivation for the work, provide background about the problem, and supply sufficient technical details and experimental results.

On the other hand, if the goal of the presentation is to get feedback to help the author improve the author's current research, he or she should plan for a very interactive style, with invitations for questions throughout. In a presentation given at a conference, questions during the talk occur but are usually clarification and are unlikely to be helpful in the sense of the previous sentence. In an invited talk at a university or company, the author may wish to encourage questions but is more likely to be met with polite silence. The audience members are present to seek stimulus that will assist them to generate their own ideas and not pass on research secrets to the presenter.

A good way to determine the message a presenter will give is for the presenter to explain his or her ideas verbally to someone who does not already understand them. This should be done before the slides have been created. This process may need to be repeated several times before the most effective way to present the material emerges. Notes should be made of the points that were made and in what order, and the presentation can be made using the notes as a base, remembering that the slides should support the talk that the scientist or engineer wants to give.

Many novice scientists and engineers (and even mature scientists and engineers) try to fit too much material into a presentation. About one slide per one or two minutes is a good pace. This rate of slide use may help the presenter to (1) focus on the key points and (2) removes the temptation to present more information than the audience can assimilate.

If the presenter attempts to fit the entire technical content of a paper into a presentation, there is the tendency for the presentation to be rushed and the audience may leave the room understanding nothing.

It is better to think of the presentation as an advertisement for the paper that gives the key ideas, intuitions, and results, and that makes the audience eager to read the paper or to talk with the presenter to learn more. The presenter may find that he or she omits entire segments of the research that do not directly contribute to the main point of the presentation.

There is always the temptation to include extra slides—the presenter should resist such forms of temptation. However, just as there should be no extra slides, there should be no missing slides.

If there is an important point to make, there should be a slide to support it. As an example, if the presenter has a new process concept (proven in the laboratory and/or at the pilot scale but protected by a patent or patents) it is wise to present a schematic of the flow lines. As another example, the presenter should not dwell on the title slide for very long, but should present a picture or text relevant to the process concept to make the motivation for his or her work more definitive and understandable.

9.2 SUBJECT MATTER

There are inherent differences between a written paper and an oral presentation of that paper, and those scientists and engineers who do not recognize this will not perform well in front of an audience. The listening audience does not behave in the same way as the reading audience. The individuals may be the same, the title of the paper and of the talk may be the same, but everything else is different because the presenter is there, and the presenter is in charge.

When the author is not there, readers are in charge and they choose their own paths through the paper—they can put it down and return to it later or read parts of it repeatedly for better understanding.

When preparing a presentation, the author should focus on the key points that he or she wants the audience to take away from the presentation. All issues or words that do not support those points should be omitted. If the presenter tries to include too much (a tempting mistake), the main points will not be assimilated by the audience, who may feel that their valuable time has been wasted. In particular, the presenter should not try to include all the details from a technical paper that describes his or her work—there are different levels of detail and a different presentation style is appropriate for each audience.

During presentation, the audience is immobile—this is not always the case if the presentation is really bad—and not free to choose any other path than the preset slide sequence. The audience cannot revisit past slides until the *question and answer period*. The audience has no control over the environment apart from the decision of where to sit.

Conversely, in an oral presentation, the author/presenter is in control: Some scientists and engineers see that responsibility as a disadvantage and would prefer to engage the audience one-on-one as in an academic conversation instead of a seemingly uncomfortable twenty-to-thirty minute monologue such as the often more uncomfortable question and answer period.

Other scientists and engineers consider the responsibility of an oral presentation as a great opportunity to have a greater influence on their reading audience and, there and then, in the seclusion of the meeting room, achieve things that many authors see only in their imagination and wildest dreams.

The failure of an oral presentation is mainly the failure of the presenter to interest the audience in *what* he or she is doing, and in *who* he or she really is. This failure often comes because the presenter attempts to do more than simply interest the members of the audience by delivering a deluge of information that could have carried Noah, his family, and the ship full of animals to Mount Ararat.

In addition, the unwillingness of the presenter to take control over the contents means that the paper is taking control. It imposes its structure, and its compressed contents are funneled through the slides. The unwillingness of the presenter to take control over time means that time is taking control with the result that the presenter, pressed for time, hurriedly presents the slides.

On the other hand, if the expectations of the presenter had been lowered to interest (and even fascinate) the audience, time and contents would no longer be so constraining.

Indeed, the failure of an oral presentation is not only the failure of the presenter to stimulate and maintain the interest of the audience but also the failure of the presenter to connect with the audience. The presentation may have been fine, but the presenter did nothing to connect to the people in the audience who might have a determining influence on the rest of his or her life.

Thus, it is not surprising that many scientists and engineers have a fear of speaking in public and it is relatively easy for such persons to find reasons why they are unable to create a technical presentation to share with others.

Yet there are many times during a career when a scientist or engineer will be asked to give a talk to a peer group or to a management group and this is a prerequisite for the scientist or engineer (Li, 1999). The *talk*, that is, the *technical presentation*, whether at a conference, to the research group, or as an invited speaker at another university or research laboratory or to a management group must be first-class and understandable to the audience (Sullivan and Wircenski, 2002; Hall, 2009).

It is during this initial time period that the subject matter of the presentation should be considered, decided upon, and finalized (Karten, 2010). Pontificating on the nature and subject matter of the talk will only lead to disaster and embarrassment.

As a prelude to this talk, the scientist or engineer should consider any presentations he or she has attended in the past, especially if they are similar in some way to the forthcoming presentation. Thoughts such as what was interesting about the other presentations and what was learned from them should be foremost in the would-be presenter's mind.

The first task in preparing a technical presentation is for the scientist or engineer to select technology that is known to him/her.

Many mature and well-seasoned professionals can confidently step before a crowded room and discuss most any topic given to them, with great poise and even an air of authority. However, for the less mature and less seasoned scientist or engineer attempting to make a technical presentation, it is better to focus on a technology with which they have confidence and experience. If there is the slightest doubt in the mind of the would-be speaker that the technology is beyond his or her technical expertise it is better that he or she forego the pain (there will be little if any pleasure) of such a presentation.

Preparing a technical presentation takes time and many scientists and engineers feel that they have to spend an inordinate amount of time doing the required work-related activities. Thus, it may be viewed as an unnecessary activity but, in reality, creating a technical presentation is always a complementary activity to the actual research/laboratory work. It should be looked upon as an activity that is designed to help the scientist or engineer to perform better in specific job-related activities as well as to expand individual careers.

A good way to determine what a presentation should include is to explain the ideas and concepts verbally to someone who does not already understand them. In fact, this step may be performed before the slides are prepared. Notice should be taken of the points that are made, and the talk should be organized on that basis. Slides should not be a crutch that constrains the presentation, but they should support the presentation that the speaker wishes to make.

The goal of a talk is similar to the goal of a scientific or engineering paper and requires that the presenter convince the audience that the research is worthwhile (is useful, solves a real problem), that it is hard (not already solved, and there are no other ways to achieve equally good results), and that he or she (the presenter) has pursued the problem diligently and has solved it.

If any of these items are not stressed or addressed to the satisfaction of the audience, the presentation is much less likely to be a success (Hall, 2009).

When giving a talk or making a presentation, the scientist or engineer should decide the key points that he or she wishes the audience to take away from the talk. Then, he or she should avoid including material that does not support those points. If he or she tried to say too much (a tempting mistake), then the main points of the presentation will not strike home and the audience will feel that the time spent attending the talk and listening to the presenter has been wasted.

A presenter should not try to do too much during the presentation. Approximately one-to-two minutes per slide is a good pace (possibly less if many of the slides are one-liners or animations that take only moments to present).

Many junior scientists and engineers make the mistake of attempting to fit a recently published technical paper into a talk causing him/her to hurry through the presentation with the result that the audience may come away hearing very little and understanding nothing. In such a case (where the talk is based on a technical publication) it is advisable to consider the talk as an advertisement for the paper that gives the key ideas, intuitions, and results, and that makes the audience eager to read the paper or to talk with the author to learn more.

However, important details should not be omitted as a result of culling too much information, but the less important details should be culled. Entire portions of the research that do not directly contribute to the main point the speaker is trying to make in the talk may also be omitted.

Just as there should be no extra slides, there should be no missing slides. As a general rule, the presenter should not speak for more than a minute or two (hence one slide every one or two minutes) without having new information appear. If the scientist or engineer has an important point to make, there should be a slide to support the point.

It is only the skilled, worldly, and effective presenters who can mesmerize an audience on a scientific or engineering topic, and leave the audience with a deep understanding of the key points, without any visual support. As a particularly poignant example, the presenter is well advised not to discuss, for example, process options having a slide or even several slides to show the flow from feedstock to products. As another example, the presenter should not dwell on the title slide for very long (again, a reminder of the one or two minutes time frame per slide) but should present a picture or text relevant to the issue under discussion.

Delivering a presentation puts the scientist or engineer in front of people he or she probably does not know. Sharing knowledge and experiences with such an audience will help them to feel as if they know the speaker and will likely help them to feel a certain sense of appreciation for the efforts the speaker has taken to create and present the information to them for their benefit.

The presenter should not present more information than the audience can assimilate. This presenter should have decided on the key points of the talk and support only those verbally *and* visually.

In summary, the presenter should formulate the goal in giving the talk. When presenting to his or her own research group, he or she should be sure to leave lots of time for discussion and feedback at the end, and to present the material in a way that invites interaction after and perhaps during the talk. In addition, when presenting to one's own group, there is less need for introductory material and any such material should go quickly for that audience but it is always good to practice giving the motivation, context, background, and big ideas.

9.3   AUDIENCE—TECHNICAL AND NONTECHNICAL

One of the first things the scientist or engineer must consider is when, where, and how he or she will make the presentation and the means by which the presentation will be made (Table 9.1). If he or she is a seasoned presenter and is comfortable with public speaking, almost any venue is suitable.

For those scientists or engineers who have less experience, they may prefer a less stressful environment, such as a local branch of the relevant technical society or

TABLE 9.1

Tips for Presenters

Audio-visual aids/equipment

Video projectors and laptops are not always compatible and computers are temperamental. Arrive early at the presentation site to ensure that all necessary equipment is set up and in good working order (this includes felt tip pens for use with a flip chart or white board)—if possible, have an emergency backup system.

Live demonstrations

Live demonstrations are risky—sometimes it is worth this risk; many times it is not.

Handouts

Handouts are not usually required for scientific and engineering meeting but may be required for internal company presentations. If required, have handouts ready and give them out at the appropriate time. It is courteous to advise the audience ahead of time that handouts (hard copies of the slides) will be available so that they will not waste time taking unnecessary notes during the presentation.

Body Language

Standing, walking, or moving about with suitable hand gestures or facial expressions is preferred to sitting down or standing still with head down and reading from a prepared speech.

Do not read from notes

It is quite acceptable to glance at the notes infrequently but continuous reading from notes shows a lack of confidence.

Maintain eye contact with the audience

Have direct eye contact with a number of people in the audience in different parts of the room and also glance at the whole audience while speaking.

Pause

The presenter should allow time to reflect and think. Racing through the presentation will leave the audience frustrated.

Humor

Use humor only when appropriate insofar as it can keep the audience interested. To do this, the presenter should know his or her audience and the humor must be on topic and not unrelated to the topic at hand.

Ending the presentation

Know when to stop talking—continuing beyond the allotted time can bore the audience especially if repetitive or unnecessary words have been used. End the presentation with easy-to-understand conclusions or with an interesting/appropriate remark or punch line.

Appreciation

Thank the audience for their time and attention—do not grovel—ask for questions or allow the session chair to ask for any questions.

---

even a company brown-bag-lunch-and-learning session. Starting in groups where the speaker feels more comfortable will help the young (or even mature socially challenged) scientist or engineer to build confidence and gain experience before feeling that he or she is being thrust into the more public speaking opportunities.

It is also important for the speaker to consider the intended audience before creating the presentation. Knowing the expectations, backgrounds, experiences, and technical leaning or preference of the audience members will help to create the most applicable presentation for them and hence give the speaker a better chance of having it well received.

In terms of getting to know the audience, the speaker may already know many of the members of small divisional groups of a technical society. However, in larger groups or conferences, knowing who may attend ahead of time is almost impossible, unless it is a specific technical symposium. There may also be session abstracts that clearly state what the goals for the session are along with any prerequisites for getting the most out of the session, and this may help the speaker control who will attend the session.

Even seasoned presenters do not always *read* their audience. Reading the audience starts before fingers are put to keyboard to commence formulating the content of slides.

The presenter must know the nature and technical caliber of the audience; otherwise there will be no presenter–audience interaction. It is immaterial whether the speaker is communicating in a one-on-one situation or to a large group; the ability to connect and understand what the audience is feeling is a critical skill for any professional who needs to persuade others.

One of the most significant signs that a presenter is starting to lose the audience connection is from body language. Such signs include but may not be limited to the members of the audience: (1) making eye contact with the speaker or looking around the room, (2) talking among themselves, (3) fiddling with a phone, a Blackberry, or a Droid, (4) shuffling through papers, and/or (5) slumped over in a chair about descend into slumber land. There are so many ways a speaker can read the audience but if he or she (the speaker) is not looking for them, he or she will proceed through the presentation seemingly oblivious to the fact that he or she lost the audience immediately after the introduction to the subject matter of the slides.

If the presenter is losing an audience, there are several methods to regain the audience's attention in which the speaker can: (1) move closer to the audience or raise his or her voice, and/or (2) become engaged and passionate, which will automatically produce greater vocal variety and cause the speaker's facial expressions to be more animated and interesting. This will encourage many to pay greater attention to the speaker and his or her message.

Many young scientists and engineers (and even the tried, true, and tested speakers) miss the signs that they are losing the audience because they focus on themselves and not on the audience and have never been (or have never bothered to be) trained to read the audience.

Most scientific and engineering professionals present slides in a one-way lecture-driven mode, and, as long as they hear the sound of their own voice, they really do believe that everything is in order and all is well with the world. This is particularly true with academics who love to hear the sound of their own voice and believe, as when they are in class, that they have a captive audience. This could not be further from the truth and is only self-delusional. Like the Energizer Bunny (of TV advertising fame), they just keep on going, caught up with their own communication and oblivious to the fact that the audience was lost (some may have even left the room) some time ago.

Being a great communicator requires the scientist or engineer to increase the speaker's awareness of the speaker-audience connection. One of the most effective ways of doing this is for the speaker to ask himself/herself what would cause him/

her, as an audience member, to lose interest or, conversely, what keeps him/her interested and focused.

Starting from this perspective can be extremely useful in improving the skills of the scientist and engineer as a communicator trying to make the speaker-audience connection, especially when members of various audiences are increasingly harder to engage, much less keep interested.

In terms of speaker behavior there are also several ways to maintain contact with the audience.

The presenter should not face the screen, which puts his or her back to the audience. This is an example of bad manners (it may even be considered to be insulting behavior by the audience) and prevents the presenter from getting feedback from the body language of the audience members—this can also cause difficulty in the audience hearing and understanding the presenter.

When giving a presentation, the speaker should *never* point at his or her laptop screen, which the audience cannot see. Using a laser pointer is in order, but the laser pointer tends to shake (not always due to an earthquake) but more especially if the speaker is nervous, which can be extremely distracting.

Some speakers prefer to use his or her hand, because the talk is considered to be more dynamic if he or she strides to the screen and points to the item on the screen. However, the speaker *must* touch the screen physically, or come within an inch of it. If the speaker does not touch the screen, audience members will observe the shadow of the speaker's finger, and the position of the shadow varies with the position of the audience member in the room.

Above all, the speaker must *never* get flustered and panic. There will be those individuals (possibly a set-up team) in the audience who may wish the speaker to make a spectacle of himself/herself. They can do this by attempting to stare down the speaker or by asking rapid-fire questions (the next question comes before the speaker has answered the previous question) from different points of the room.

One approach is for the speaker to stop and regroup; taking a drink of water is a good way to cover this, so the presenter should have water on hand even if he or she does not suffer from dry throat. Another approach is for the speaker to ask the questioners to have the courtesy of waiting until he or she has answered the previous question before firing in another question. Most members of the audience do not wish to see the speaker being berated and embarrassed by such events and will be supportive of the speaker when he or she attempts to restore a semblance of order.

9.4 INTRODUCTION, BODY OF THE PRESENTATION, AND CONCLUSIONS

For the very beginning of the presentation, the speaker needs to convince the audience that this talk is worth paying attention to—it is solving an important and comprehensible scientific or engineering problem. Some presenters prefer to commence with an outline slide that essentially gives the audience the overview—*this is what I will tell you.*

The up-front motivation can be useful to show an outline slide at the start of each section, to help the audience stay on track (or help those who got distracted or lost to rejoin the presentation), but often the speaker does not need one for the introductory, motivational section of the talk.

If the presentation is fairly complex insofar as several issues need to be addressed, it may be preferable to use outline slides throughout the talk: at the beginning of each major section of the talk the outline slide is shown with the current section indicated using color and an arrow. This helps the audience to regain its bearings and to keep in mind the structure of the presentation. Some presenters are against using an outline slide at the commencement of the presentation but, in reality, there is no hard and fast rule. The choice of slides is dependent upon the nature and content of the presentation and the nature of the audience.

When a subsequent slide adds material to a previous slide (or in some other way just slightly changes the previous slide; this is sometimes called a *build*), all common elements must remain in *exactly* the same position. A good way for the speaker to check this during formulation of the presentation is to quickly transition back and forth between the two slides several times. If there is any apparent *shudder*, it should be corrected—this may require that extra space be assigned on an early slide to accommodate text or figures to be inserted later; even though that space may look a little unnatural, it is better than the alternative.

Some speakers have been known to add one extra slide to a presentation previously made at the same meeting one year or six months before. At one meeting attended by the author of this book, one speaker added jokingly that this was his way of making sure he was invited back again next year. This is to be frowned upon and borders on unprofessional behavior. The audience may not be amused!

When the audience recognizes that something is different, they are generally uneasy about exactly what has changed (the human eye is good at detecting the change but only good at localizing changes when those changes are small). The speaker wants the audience to have confidence that most parts of the slide have not changed, and the only effective way to do that is not to change those parts whatsoever. You should also consider emphasizing (say, with color or highlighting) what has been added on each slide.

Slide content is as important as slide layout (surprise)! A slide should not contain too much text—an example of the amount of text is *no more than 75 characters on a slide*. Some presenters give the impression that they have written a novel and the content is all on one slide.

When a new slide is shown to the audience, the members turn their attention to comprehending the slide. If the audience has to read a lot of text, they will usually sever the presenter-audience contact and may even miss an important point. This is not the fault of the audience but it is the fault of the speaker.

As a general rule of thumb, three lines of text per bullet point are usually too much, and two full lines of text per bullet point is usually bordering on more than enough (this depends upon the size of the font that has been used to make the slide readable to all in the room). If there is too much text, it should be shortened or broken it into pieces (by using sub-bullet points) so that the audience can read the text quickly without having to ignore the presenter.

The speaker should not read the slides word-for-word. Any speaker who has to do this has used too much text on the slides. It must be remembered that the slides are an outline of the work and not a transcript. Reading the slides verbatim is very boring and the speaker will sever any speaker-audience bond that has been created.

As already stated, fonts should be large and easy to read from the back of the room. If something is not important enough for the audience to be able to read, then it probably does not belong on a slide.

However, caution is advised in slide design because it is possible to overdo the practice of limiting what information appears on each slide. The presenter needs to show sufficient (there may never be enough!) material to support the work presented—especially if there are questions or to show that the simplified version of the work contained in the presentation is an accurate generalization of the work. But the mistake of including too much information is a far more common error exhibited by all (seasoned and unseasoned) presenters.

Throughout the presentation, the presenter should make eye contact with the audience. Not only does this ensure that the audience is interested (if they know the speaker is looking at them, the tendency to sleep may be resisted) but it also allows the presenter to know whether he or she is talking too quickly, too slowly, or at an acceptable pace (sometimes referred to as the *Goldilocks syndrome*).

9.5   NUMBER OF SLIDES

The number of the slides for any presentation will vary with the background and expertise of the audience. However, the concept that there should be approximately two minutes of talk per slide is reasonable and workable, but there should never be less than one minute per slide. On the other hand, for a class lecture to students, it is permissible to have a large number of slides.

The slides should have descriptive slide titles and the same title should not be used on multiple slides (except perhaps when the slides constitute an animation). The title should help the audience to focus on the slide and also appreciate the specific contribution of the slide. If the audience are left dumbfounded by any one or more slides, it not only suggests but proves that the speaker has not done a good job of understanding and organizing his or her own material. If this is the case, the question period will be very interesting for the speaker!

The last slide should be a contributions or conclusions slide, reminding the audience of the take-home message of the talk. Some speakers will end a talk with a *future work* slide, or with a slide that indicates time for *questions* or *thank you* or *the end* or merely gives the speaker's e-mail address. This is a personal preference of the speaker and the forum at which the talk is presented. Some speakers leave the conclusion slide up after he or she has finished the talk (while questions are being answered).

One way to consider the content of the *last slide* is decide what the speaker would want to be the last thing that the audience sees (or that the audience sees while the speaker fields questions).

9.6   CHARACTER OF THE SLIDES

The presentation should commence with motivation and examples—and should have motivation and examples throughout. The presentation usually follows a specific format: (1) introduction, (2) description of the work, and (3) conclusions.

First and foremost, each slide should have an individual and descriptive title. Use of the same title on multiple slides (except perhaps when the slides constitute an animation or build) is often confusing to the audience, and is particularly confusing during the question period when an audience member refers to a slide by its title. The title should be chosen to assist the audience to appreciate the specific contribution of the slide. Also, the slides should be individually numbered, usually in the lower right-hand corner for ease of reference by the presenter and by the audience.

The *initial slides* should be introductory with examples. The audience needs to be convinced that the presentation is worth their attention as it relates to solving an important technical problem. The first slide (after the title slide) should be an example of the problem to be solved or some other form of motivation that will catch the attention of the audience.

Outline slides can be useful, especially in a talk that runs longer than thirty minutes, because they help the audience to keep in mind the structure of the presentation and the problem being addressed. An outline slide (with the current section indicated by means of color, a change in font, an italicized font, and/or an arrow) can be used at the beginning of each major section of the talk, other than the introductory (or attention-catching) section.

When a subsequent slide (in the heart of the presentation) adds material to a previous one, all common elements must remain in *exactly* the same position. A good way to check this is to quickly transition back and forth between the two slides several times. The transition should be smooth and not exhibit any change in format or line positions. The audience should have confidence that most parts of the slide have not changed, and the only effective way to do that is not to change those parts whatsoever. It may also be appropriate to consider emphasizing (with color or highlighting) what has been added on each slide.

The slides should remain uncluttered—when a new slide is shown on the screen, the audience will focus on comprehending that slide. If the audience has to read a lot of text, they may miss an important point—diagrams must be simple and clear and, as a general rule, two full lines of text for a bullet point is moving toward a cluttered slide. In such cases, the text should be shortened or broken into pieces (sublevel bullet points are acceptable) so that the audience can skim though the slide without having to ignore the presenter for too long.

The *final slide* should be a contributions or conclusions slide, reminding the audience of the take-home message of the presentation. This slide can remain in view after the presentation has been completed while questions are being answered. This will be the last item that is viewed by the audience.

Generally, once the presenter has decided on the number and format of the slides, there are other general rules to be observed. Thus, the presenter should do the following:

- Not read the slides word-for-word—reading the slides verbatim will cause the audience to lose interest.
- Use the slides as an outline, not a transcript of the presentation.
- Use the slides so as to give the main points—more detail can be supplied verbally.
- Use the slides as a base from which to speak and to remember all the main points and the order in which they should be presented to the audience.
- Use the diagrams as backup to the presentation in place of extra words and to explain whatever is important, interesting, or novel about the work.

It is possible to overdo the practice of limiting the information that appears on each slide, and there should be sufficient information to support responses to questions or to show that a simplified model presented verbally is an accurate generalization.

Fonts should be large and easy to read from the back of the room. Including *fine print* (as happens in many legal contracts) means that it probably does not belong on the slide.

The presenter should make effective uses of figures and, if possible, needs to avoid a presentation that is just text. Such a presentation may miss important opportunities to convey information. Images and visualizations are extremely helpful to most audiences. However, pictures and text are better than text alone, and text alone is better than text plus bad pictures.

Images and visualizations are extremely helpful to the audience. The presenter must make effective use of figures and color, all diagrams must be simple and visually clear, and the title must be telegraphic. The speaker should avoid a presentation that is (literally) just dozens of pages of text. Also, the speaker should know that a number of the audience may be color-blind (approximately 5% of American males are color blind) and the color should be augmented with other emphasis wherever possible.

The presenter should include diagrams to show how a process works or how the process reactor is integrated with other units or reactors. But, it should not be overdone and strive to include images just because images are usually a good idea.

Generic images that do not relate directly to the presentation should be omitted. For example, if there is a slide about refinery distillation, the generic image of a kettle on the stove boiling water (which is a form of distillation) to make tea is totally unwarranted. Good meaningful diagrams or pictures and text are better than text alone but text alone is better than text plus bad diagrams or pictures.

When a slide is used that includes a diagram, background should be the same color as that of the slide. For example, if the slides have a blue background, using a diagram with a white background can be visually distracting to the audience as it may well be hard to read, and unattractive.

One might often hear the term *eye candy* in reference to presentations. The term usually refers to transition effects, design elements that appear on every slide, or multicolor backgrounds. At best, the audience may not be amused and maintain an air of dignified distraction or, more likely, the audience will be distracted and the points that the speaker wishes to make will be lost.

At one meeting attended by the author of this book, one presenter seemed to have a fascination with the color yellow and designed many slides (too many— he had 100 slides for a 60-minute presentation) and each slide was packed with information—approximately 150–200 characters of white text on the yellow background per slide (yellow text on a white background can be equally confusing and offensive to the audience). It may come as no surprise that members of the audience were alienated—many had the impression that the speaker was more interested in graphical glitz and over-verbose text than in content. His slides were unattractive and unreadable. He did not ensure that each element on the slide contributed to his message.

There is a variety of fonts that can be used for slides. Most seasoned presenters have learned to prefer a sans-serif font for their slides. While serifed fonts are satisfactory for reading on paper, sans-serif fonts are easier to read on a screen.

Finally, many presenters prepare their slides with a **bold** font, with the use of color, or underlining for emphasis where necessary.

Above all, the slides should not distract or alienate the audience from the message to be presented. The slides should be attractive and compelling without being fancy. Make sure that each element on the slides contributes to the message; if it does not, then remove it.

9.7   QUESTION PERIOD AND BEYOND

It is customary to end a presentation with a short period for questions and even the seasoned scientist or engineer who has made hundreds of presentations can find the question period to be daunting. Answering questions from the audience requires skill and the ability not to appear stumped, dumbfounded, or angry at the questioner!

Even after a scientist or engineer becomes very proficient at giving a talk, it will probably take the speaker quite a bit longer to become good at answering questions. It requires work to improve this aspect of the presentation.

The presenter can expect to receive three types of questions: (1) the genuine request for knowledge, which should cause the speaker no difficulties if he or she is adequately prepared; (2) the self-centered question, in which the questioner merely wishes to draw attention to him or herself and elicit wonder from the audience at his or her ability to devise such an incisive and cogent question; and (3) the malicious question, in which the questioner attempts to expose the speaker as a know-nothing person.

The best defense against any type of question is to be well versed in the subject of the presentation; to be polite, forthright, and honest; and overall to avoid getting involved in a lengthy exchange. It is sometimes advisable to short-circuit such questions by offering a one-on-one discussion after the presentation or after the session in which the presentation was made.

Therefore, just as the would-be presenter practices the verbal part of the presentation, there should also be practice at answering questions—both the predictable questions and the unpredictable questions. Giving practice talks to other people who are willing to ask such questions can be very helpful.

Answering questions from the audience can be difficult, to say the least! Even after a scientist or engineer becomes proficient at giving a talk, it will probably take him/her longer to become proficient at answering questions.

When an audience member asks a question, it is a good idea for the presenter to repeat the question, asking the questioner whether he or she have understood it, before answering. This has three benefits:

- The speaker ensures that he or she has understood the question. When thinking under pressure, it can be far too easy to jump to conclusions, and it is unprofessional to answer a question different from the one that was asked. A related benefit is that the speaker is able to frame the question in his or her own words or from his or her own viewpoint.
- The speaker gives himself/herself a few moments to think about the answer.
- If the audience member does not have a microphone, the rest of the audience may not have been able to hear the question clearly.

Above all, do not pontificate when answering a question. That only serves to alienate the audience. The speaker should be willing to answer a question with *no* or *I do not know*. Pontificators get into more trouble when they answer a question with a *book-chapter of words*. This only serves to emphasize that the answer should have been *I do not know*—with the added mental or verbal response *but I will find out*.

Above all, do not confuse *I do not know* with *it is not known*. The presenter should only respond in the latter form when he or she is sure that the question is open. If the presenter has to respond with *I do not know*, the response should be stated with assurance and confidence rather than with a meek or timid answer (Parberry, 1989).

Beyond the question period, talking with attendees after the presentation also affords the speaker an opportunity to solicit feedback from them—particularly about their likes and dislikes regarding the presentation. Giving audience members the time and permission to speak candidly will enable the speaker to improve upon it for the next time. If evaluation forms were collected, the speaker should try to gain access to them so he or she can determine how well or how badly the presentation was assessed.

While the presentation is still fresh in the speaker's mind, it is always a good idea to update the presentation for the next time. This could involve making adjustments to individual slides and adding or removing slides as needed. The speaker may also wish to revise the script, capturing important points that were not apparent beforehand but that became obvious during the presentation.

9.8 PRACTICE TALK

Last but certainly not least, when all of the preliminary and planning work has been done, the presenter is well advised to seek feedback on his or her work by giving a practice talk. One of the most effective ways to improve a presentation is to see the reactions of others and get their ideas and advice.

As scientists and engineers gain experience making technical presentations, they will become more and more proficient at it and may even be able to pull a presentation out of their repertoire and deliver it with a moment's notice. However, until that point is reached, it is best to be well prepared. The speaker should go through the presentation until he or she knows the order of the slides by instinct and it is always advisable to develop transition comments to help ease the audience from topic to topic and slide to slide.

The would-be speaker should also rehearse the timing of the presentation to make sure he or she knows where the halfway point is to help allocate time during the presentation. Knowing the subject matter and slides instills a sense of confidence as the speaker progresses through the presentation.

The practice talk should, if the occasion demands, be used on a regular basis by those scientist and engineers who feel they are socially challenged and unable to speak in public. The most cranially qualified scientist or engineer can be significantly hindered if he or she cannot adequately express his or her knowledge to others. Being able to succinctly take a complex concept and communicate it in a manner that is easy to understand by one's peers or to those less technical is truly a skill that is as highly sought after as the technology expertise itself.

In fact, creating technical presentations whether or not they are needed imminently is a good way to practice organizing thoughts, speaking with clarity, and communicating effectively with others.

In addition, it is also a good idea to give a practice talk before the speaker makes the presentation in front of an audience. Even if the scientist or engineer has reviewed the slides and thinks he or she knows how the talk will go, when speaking out loud one's ideas are likely to come out in a different or less clear way. Remember friend, Murphy.

Murphy's Law: Whatever can go wrong will go wrong.
Speight's Law: Murphy was an optimist!

Whatever is true about writing is also true of making a presentation: even if the presenter knows what he or she wishes to say, it takes several revisions to decide on the best way to say it.

In fact, the speaker should practice the talk to himself or herself—speaking out loud in front of a mirror, for example—before he or she gives the first practice talk. In such a practice session, the presenter must say every word he or she intends to in the actual talk, not skipping over the parts that are difficult.

A tape recorder is also a handy instrument—it will help the speaker decide if his or her timing is right. In addition, a videotaping of the solo presentation will enable the speaker to see how he or she appears to others—this information can be somewhat traumatic, but it is invaluable in helping the speaker to improve.

It may be a good idea to keep the practice talk audience relatively small—to less than ten people. In a large group, many of the listeners may not bother to speak up, and if the pool of potential attendees is large, it gives the speaker the chance to give multiple practice talks, since the best feedback is given by someone who has not seen the talk (or even the material) before. Giving multiple practice talks is essential for high-profile talks such as conference presentations.

However, the group should not be too small, otherwise the speaker might be convinced to change the entire structure of the talk by one person who has a different view. Getting a balance of opinions will help the speaker avoid making too many mistakes.

When giving a practice talk, it is advisable to number the slides (say, in the bottom right-hand corner), even if the speaker does not intend to include slide numbers in the final presentation.

In addition, it is very helpful to distribute hardcopy slides (with slide numbers included) so that the audience can easily annotate them and return them at the end of the talk. Also, the audience will spend less time trying to describe what slide their comment applies to, and more time writing the comment and paying attention to the speaker. For nonpractice talks, it is not general practice to give out hardcopies of the slides, as they will tempt the audience to pay attention to the piece of paper instead of to the speaker.

Finally, when possible, all would-be presenters should attend other practice talks—cultivating these obligations is a good way to ensure that the would-be speaker has an audience at his or her practice talk. Furthermore, attending practice presentations by other scientists and engineers can teach a young scientist and engineer a great deal about good and bad talks—both from observing the speaker and thinking about how the talk can be better (or is already excellent) and from comparing the feedback of audience members to personal opinions and observations.

REFERENCES

Alley, M. 2003. *The Craft of Scientific Presentations*. Springer-Verlag, New York.

Hall, R. 2009. *Brilliant Presentation Revised: What the Best Presenters Know, Do, and Say*. Trans-Atlantic Publications, Philadelphia, Pennsylvania.

Karten, N. 2010. *Presentation Skills for Technical Professionals: Achieving Excellence*. IT Governance Publishing, Ely, Cambridgeshire, United Kingdom.

Li, V.O.K. 1999. Hints on Writing Technical Papers and Making Presentations. *IEEE Transactions on Education*, 42(2): 134–137.

Mandel, S. 1994. *Technical Presentation Skills*, Revised Edition. Mandel Communication, Capitola, California.

Parberry, I. 1989. How to Present a Paper in Theoretical Computer Science: A Speaker's Guide for Students. *Bulletin of the European Association for Theoretical Computer Science*, 37: 344–349.

Sullivan, R.L, and Wircenski, J. 2002. *Technical Presentation Workbook: Winning Strategies for Effective Public Speaking*. ASME Press, American Society for Mechanical Engineers, New York.

Walker, T.J. 2010. *How to Give Pretty Good Presentation: A Speaking Guide for the Rest of Us*. John Wiley & Sons, Hoboken, New Jersey.

10 Other Forms of Writing and Presentation

10.1 INTRODUCTION

In addition to the typical writing and presentation efforts by scientist and engineers (Chapter 2 and Chapter 9), there are other forms of writing and presentation that need to be taken into account.

Conveying research findings is not exclusively limited to publication of papers, writing a technical report, (Chapter 2), or making a formal presentation (Chapter 9). There are two other forms of presentation that a scientist or engineer can use. These are (1) technical writing in a trade magazine and (2) a poster presentation at a scientific or engineering meeting.

The myriad forms of writing used by scientists and engineers are often referred to as technical writing or *prose writing*. However, these terms are so general to the point of being ambiguous that some qualification or clarification must be made when referencing such work.

The different forms of writing that a scientist or engineer might have to deal with at some point in his or her career include (1) technical writing (i.e., writing for trade journals), (2) writing for nontechnical publications, and (3) poster presentations. While there are undoubtedly other forms of writing for scientists and engineers, these are the most common forms.

While the qualities of good writing cut across all genres of writing, each form of writing may have individual rules—the form of a letter is different from a non-technical article, for example.) In each case, a scientist or engineer must take a clear stand on an issue and anticipate all objections or contrary arguments to his or her point of view.

Technical writing is the writing about scientific and engineering technology in which the information is put into simple words so that users of the technology will understand the new technology and can use it freely. In *prose writing* (sometimes called *literary writing*), which is completely different from technical writing, the writer tends to focus on the human world.

The term *poster presentation* as used in the context of this book refers to posters presented at scientific or engineering conferences that illustrate the results of scientific or engineering research.

On the other hand, while the term *technical writing* incorporates writing scientific and engineering papers and reports, it is used in the context of the present chapter to refer to publications in trade journals and in nontechnical publications, such as newspapers and the like. This involves promoting science and engineering concepts in magazines not usually read by the scientific or engineering specialist. As always, the writer must consider three principles: (1) the purpose of the article, (2) the makeup of the audience, and (3) the manner in which the information is presented.

Most technical writers have expertise in one area, and specialize in that. Technical writing requires great knowledge of the scientific or engineering subject and is usually aimed at people who also understand it, as in a scientific process or an engineering problem. Technical writing is not like writing literature because it does not require broad appeal but instead is aimed at a specific audience. It also does not use figurative language, but is instead straightforward and informative.

Another document the scientist or engineer may be called upon to write is a business *white paper*, which is a report to educate readers (often upper management or a Board of Directors) and which can also function as a marketing tool. For example, the author of a *white paper* may present an informative look at a problem that can be solved by the company's product. The business white paper tends to be less straightforward and longer than a technical brochure. Scientific and engineering writers who are not necessarily well-versed in the product or topic can interview experts for in-depth information.

Thus, in all cases, the scientist or engineer must have a clear understanding of the purpose of the document he or she will create (Lamott, 1994). Such writing, like any other form of scientific or engineering writing, typically requires information gathering from existing documentation and from subject matter expert sources. Scientists and engineers at many levels and in many different subdisciplines have a role in producing technical communications. And, as in all forms of scientific and engineering writing, strong language skills and (in some cases) teaching skills are essential and the author must understand the many conventions of modern technical communications.

In all cases, clarity in reporting how the research was conducted and what results were obtained is paramount. Again, only through clear and thorough presentation of ideas can other scientists and engineers transfer the benefit of the research.

10.2   WRITING FOR TRADE JOURNALS

A trade magazine (in the context of this book) is, as the name implies, about the business of science and technology. Technical writing for a trade magazine concerns a technical topic, and it is brief, logical, and explanatory, in basic or simplified understandable language (Hanson, 1999).

Briefly, trade magazines are periodicals that are published for and read by members of specific trade groups, occupations, and/or persons involved in particular types of businesses. Such magazines are not always available to the general public—most are available by subscription only, or as a premium for membership in an organization or association. They range from pictorial presentations to modest, staple-bound publications.

Furthermore, trade magazines are not professional journals in the usual sense of the term—while professional journals (such as *The Journal of the American Chemical Society* or *The Journal of the American Institute of Chemical Engineers*) rely heavily on theoretical content and scientific or engineering details, trade magazines focus on the basics and short articles.

Technical writing for a trade magazine is probably the most widely read form of written scientific and engineering communication, with the exception of advertising. Businesses use technical writing extensively to communicate information to management, employees, and other businesses. Technical writing in business can include training manuals, employee guidelines or handbooks, and other specialized writing that is task specific. On the other hand, technical writing geared to the general public is only one small aspect of the field.

As with all forms of writing, to write effectively for a trade magazine the writer needs to find a physical space where he or she can both work and think. For many such writers this space is not always the same office he or she uses for writing the typical paper or report—the thinking paradigm is changed and the physical surroundings may need to change for maximum productivity. The course of everyday business and the influence of the everyday business environment may not be conducive to the new form of thinking.

Scientific and engineering writers who write for trade magazines address the larger semi-technical and nontechnical readers about the science and technology advances. In such cases, the writing is very different to that used in the scientific or engineering paper (Chapter 2). Part of the difficulty in writing such articles is that the knowledge base of the audience is much more diverse, and more limited, than that of the scientific or engineering communities in general.

The readers may know something about science or engineering but not necessarily know the details of the roles played by theory and observation in developing a new concept or operating a process unit. Hence, science and engineering writers for trade magazines (and other forms of semitechnical publications) have to *explain* much of the article content from the significance of scientific and engineering facts to the methods used while simultaneously holding the reader's attention.

The major difference between an article written for a trade journal and for a scientific or engineering journal is that there is typically no *Results* section in the trade journal article. However, the writer must ensure that the data he or she chooses to show is representative of the whole picture to avoid misleading the audience. There is no problem in selecting particular phrases and descriptions that clearly demonstrate the findings but there should not be a focus on findings that embellish the concept and move it into the realm of *Fantasy Island*. This would be misleading since the findings used by the author do not fairly represent the original observations. The article may not be truly technical but it must be honest.

Thus the purposes and audiences are different between journal-style scientific and engineering writing and technical writing for trade journals. The purposes of journal-style writing is (1) to present the results of a scientist's or engineer's knowledge, (2) to present the results gained from scientific and engineering personal research, and (3) to present the point of view or hypothesis originated and/or developed by a scientist or engineer.

Both scientific and engineering writing for a journal and for a trade magazine contain technical jargon, but the jargon is used for different purposes and the audiences are completely different. Understanding the audience for either publication medium is a must (Chapter 4).

The audience for journal-style writing is scientists and engineers involved in similar fields of research. The focus of an article in a trade magazine can be (1) to educate the audience about the benefits of using a specific product or service and/or (2) to describe the procedures that are employed by companies for carrying out various tasks. The audience will, of course, differ from product to product. In addition, technical writing differs from area to area.

Thus, trade magazines have a narrower focus than their consumer counterparts and the types of articles they carry fall into categories such as the following:

- News items specific to the industry focus of the magazine readership
- Products and trends
- Process-related articles such as how-to-optimize-process-parameters
- Personal/professional experience articles, such as case histories of process operations

The scientist or engineer who is invited to write an article for a trade magazine is usually established in his or her profession and has been noticed by the editor as a potential contributor.

The editors of trade magazines look for two types of writers: (1) one may be a vendor or manufacturer of process equipment, and (2) the other is more likely to be a consultant or independent expert. The latter may be a person who associates with personnel in a specific industry (such as at trade shows, trade organization meetings, and the occasional technical seminar) and will have the credibility and background to write the article.

Thus, whether information on technology, assembly instructions, or owner's manuals, technical writing for a trade magazine should carry a message that is clear, concise and, most importantly, helpful to the audience. The scientist or engineer who writes for a trade journal must be able to engage the intended audience in order for the readers to understand and assimilate the information.

Technical writing is arguably more difficult than other forms of writing because of the fact that it must be clear and to the point. Good technical writing should not leave any room for imagination and it must anticipate and answer any questions or problems that may arise in the mind of the reader.

Above all, the scientist or engineer who moves into technical writing must possess excellent writing skills and know how to write the article to fit the intended audience. Although scientists and engineers who write books on complex topics may not need to simplify the material too much, those who write training materials need to be able to write for users who possess less experience. Other skills needed to be a successful technical author (in the sense of the trade magazine) include fluency in office productivity software, experience with formatting documents and extensive knowledge on the technical topics in which they specialize. There is also the need to have basic editing skills, especially in cases when the scientific or engineering author is solely responsible for the final work.

10.2.1 Getting Started

Getting started on an article for a trade magazine is often the most difficult part of the task.

An article in a trade journal may be written by a single author only, but multiple authors are also featured on many articles. As with a journal-type article there should be a corresponding author (Chapter 5).

As a brief recap, the use of a corresponding author prevents multiple authors from slowing down the review and publication process. Any questions or comments the publication may have regarding the submitted work can be directed to one author, who may then, after discussion with his or her coauthors, provide a single, authoritative response. The corresponding author is typically responsible for reviewing any drafts and changes sent by the publisher. This is why the choice of author to act as the communication point of the group of coauthors is important—any errors or incorrect changes that go to print are the responsibility of this individual.

If the scientist or engineer does not receive copies of the magazine through membership in the association, he or she should begin by researching past issues of the magazine—an analysis of at least six months of back issues of the magazine (this would not be out of place) *and* a copy of their guidelines for authors.

Articles that appear in trade magazines are different in style than those seen in professional journals. The writer should pay attention to (1) the sentence structure, (2) commonly used terms, and (3) the assumed knowledge of the readers.

In order to write a meaningful and successful article for a trade journal, the scientist or engineer should (1) be well-versed in the subject matter of the article, (2) know the magazine, (3) know the aim of the magazine, and, last but by no means least, (4) be up-to-date on industry news.

Being well versed in the subject matter of the article is essential. In order to write for a trade magazine, the scientist or engineer will need and in-depth knowledge of the topic to be covered. Trade magazines usually require professional knowledge of a topic, so the writer should make sure it is a subject he or she will want to spend time on. Some trade magazines accept articles of personal experience or interviews with recognized authorities in their field.

In addition, the scientific or engineering writer needs to be able to convey information using the *language* of the readership. Many trade magazines use technical terms that are a foreign language to industry outsiders. This may involve translation from the technical language (or jargon) of the writer's discipline to the technical language (or jargon) of the reader. If the writer finds that he or she is not as knowledgeable about the topic as would be preferred, there are two options: (1) gracefully decline the editor's invitation giving reasons, (2) accept the editor's invitation and use as many resources as are available, including other people, to learn more and more about the topic.

The writer should also *know the magazine*. He or she should be (at least a reader but preferably) an avid reader of the magazine so that he or she is adept at following the same writing style. In addition, the writer should know the types of articles that appear in each issue.

The writer should also *know the aim of the magazine*, which is typically to provide the readers with information that increases the integrity of industry by fostering shared knowledge. The content of the article should be focused and varied to maintain interest of the readership.

The writer should also be up-to-date on industry news that affects the field being described in the article. The best scientific and engineering writers will be abreast of the latest news, making them more apt to write appealing and innovative stories for the magazine.

However, if articles in trade journals are to be really useful and appealing, readers must be able to understand and employ them without having to decode wordy and ambiguous prose. For example, the engineer who wishes to read about, say, the *chemistry of petroleum refining* may prefer to see the chemistry described in simple terms without complex formulas and descriptions of multistep kinetics equations—these can be accommodated by meaningful citation in the article.

Good technical writing clarifies technical jargon; that is, it presents useful information that is clear and easy to understand for the intended audience. Poor technical writing may increase confusion by creating unnecessary technical jargon, or failing to explain unavoidable technical terms of which the readers have little or no knowledge.

As with any scientific or engineering publication (or presentation) there are several key issues that must be considered, namely the following:

- The audience: Is the audience composed of people who have some experience and knowledge of science or engineering or is the audience new to such technical subjects?
- Writing style: What is the style of writing and are there any current available issues of the magazine that illustrate the acceptable writing style?
- Deliverable: Is the deliverable a simple text for inclusion in a book, a magazine, a newspaper, or a Web page?

Whatever the responses to the above questions, the writing must be (1) clear, (2) concise, (3) complete, and (4) understandable. Clear, concise, complete, and understandable technical writing will help the reader to grasp the meaning quickly and it will be retained within the reader's memory.

Audience analysis (Chapter 4) is a key feature of all technical writing. Technical writing, like all forms of scientific and engineering writing, is a means of communication in order to convey a particular piece of information to a particular audience for a particular purpose. Thus, to present appropriate information, writers must understand the audience and their goals and may even wish to use illustrations to make the benefits easier for the audience to understand.

In fact, technical writing involves an attractive layout for easy reading and comprehension. Strategies used in technical presentations will help readers to grasp messages quickly use one or more of the following:

- The top-down strategy: tell the audience what the scientist or engineer will say, say it, then tell the audience what the scientist or engineer has said
- Headings (like headlines in newspapers)
- Short paragraphs
- A plain, objective style so that readers of all disciplines can easily grasp details

Once the scientist or engineer has decided that he or she will submit (or has been invited to submit) an article to a trade journal, the path is not always clear of obstacles. Trade communities can be webs of intrigue insofar as the members communicate with each other. In fact, trade communities have even been called incestuous—in the communication sense of the word.

Therefore, the writer *must not* insult any of the community members or even hint of faults of any kind. The consequences can be irreversible and professionally fatal for the fledgling (even mature) trade-journal writer.

10.2.2 WRITTEN PRODUCT

The ideal article in a trade journal is one a reader preserves online or, in the old-fashioned way, as a hard copy or reprint. The benchmark by which many trade magazines are measured is the usefulness of the articles. Any self-respecting reader will ask himself/herself, "Will this help me?" If the answer is positive, all is well—at least for this edition of the journal. If the answer to that question is negative, the article (and the writer) has failed. Word will spread like the proverbial wildfire and the career of the scientist or engineer as a writer for trade journals is over.

In addition to the usefulness of the article, trade magazines have other requirements for content:

- Use of quotes is usually not always a plus in trade journals—readers are rarely thrilled by name-dropping and prefer to know what an expert says about how they can, for example, optimize their process parameters.
- Product pitches usually indicate that the writer is not objective—the credibility of the writer then become suspect.
- Trade magazines typically occupy a niche between general interest magazines and professional journals but remain as understandable and readable as the typical consumer magazine.
- Trade magazine graphics are not always easy to compile and are not always acceptable. The writer may need to discuss any such article insert with the magazine editor to determine what he or she will accept and will not accept. The editor may even offer to redraw the graphic from a PowerPoint sketch.

Once the scientist or engineer has an article published in a trade magazine, he or she may receive other invitations, and not always from the same magazine.

In summary, the written product should be (see also for other examples Table 10.1):

Clear—leaving little possibility for questions about the meaning of any fact presented.

Complete—having enough detail to support itself, while citing any external sources.

Concise—short and to the point, since the readership will not appreciate reading over-verbose documents.

Legally acceptable—any claims in the document must be truthful and within the limits of legal acceptability.

Accurate—there should be no embellishments or false advertising.

TABLE 10.1

Tips for Writing a Technical (Nonjournal) Document

Parameter	Comment
Content	Factual, straightforward, understandable
Audience	Specific within a discipline or subdiscipline
Purpose	Inform, instruct, and persuade
Style	Formal, standard, usually nonacademic
Tone	Objective and to the point
Vocabulary	Specialized but understandable for the nonspecialist
Organization	Sequential, systematic, easy to follow

While clarity, completeness, and being concise are of the essence, the challenge is to hold the interest of the reader while complying with the various legal (national and state) mandates placed on technical documents.

A technical document can suffer severe damage and loss of credibility if it contains inaccurate information. While the technical content is essential, the overall quality of documents matters a lot to the audience. If there are obvious grammatical and mechanical errors, damage to credibility is unavoidable and once a reader detects an error, he or she will, most likely, question the quality and credibility of the entire document.

Finally, as with all forms of scientific and engineering writing, the message determines the medium. The writer must focus on (1) the message to be conveyed, (2) the audience receiving the message, and (3) the most appropriate trade journal. The writer may be able to resolve these issues and may have to discuss them with a more experienced colleague.

10.3   NONTECHNICAL PUBLICATIONS

At some point in the career of every scientist or engineer there comes an invitation to write an article for (or make a presentation to) an audience that is nontechnical. The group may consist of executives, students, or members of a community organization.

Explaining scientific and engineering advances to a nontechnical audience is a worthwhile endeavor. If the scientist or engineer has a substantial background in the subject matter of the publication or presentation, he or she will have an advantage over less professionally qualified writers, who may have been trained in writing rather than in science or engineering.

However, as experts in their respective fields/disciplines, scientists and engineers tend to know their areas of expertise too well. As a result, the scientist or engineer will be so deeply immersed in their fields that others have a problem understanding them. There is the danger of writing or presenting at a level that is too detailed for the target audience, whether the audience be composed of senior management or end users, but in either case the problem is the same for the scientist or engineer and that is not being understood.

As the write/presenter structures his or her ideas, there is always the discovery that it is difficult to communicate science and engineering in simple, lay terms. It is, in fact, a daunting hurdle to translate technical information into everyday language (Matthews et al., 1996). The writer can assume that while he or she may know much more than the members of the audience, they most likely know more than the writer does about a number of other topical issues.

Thus, writing for a nontechnical audience and readership requires as much care and forethought as it takes to write for a technical audience or readership. If the writer assumes too much about the abilities and knowledge base of the audience, the results can be disastrous. The audience can be left feeling frustrated and much less likely to ever read about such a topic again.

On the other hand, the writer must not *dumb-down* the article, otherwise the audience would not be inclined to read the article. Because of some interest by the audience, the writer must also give the audience credit for being able to find further information (as required) by doing a *Google search* or some other form of research when interest in the topic and is high and worth further exploration.

Once the writer has a sense of the knowledge base of the audience, this will provide guidance about what to cover and what to omit. Selecting the appropriate scope is essential because it can seriously influence the ability of a writer to explain the topic in an effective manner. If the article is too broad, it will be too diffuse and/or too overwhelming. If the article is narrow in scope, the writers runs the risk of getting bogged down in detail or boring the audience with too little of substance.

The writer/presenter must be prepared to educate the audience on why the topic is important. Indeed, preparing an article or a presentation for a nontechnical audience requires as much forethought and work as preparing a detailed technical paper for submittal to a journal or as an internal company report. As always, knowing the audience is of the utmost importance (Chapter 4). If the author assumes too much about the audience or selects the wrong technical level to address, the results will be disastrous—the audience may actually end up being even more confused about the topic as they were before reading the article or listening to the presentation.

The author/presenter must decide the means by which he or she will define the audience as well as the relationship (if any) of that audience to the information that is being delivered. While a technical audience might crave extra detail because it is critical for their work, a nontechnical audience might not understand, or worse, not even read a document that is heavy with jargon.

Typically the audience will vary widely in their technical knowledge and level of comfort with technical topics. It is the task of the author to determine a middle ground between the extremes of audience knowledge. There will always be faults for the author (1) being too technical, or (2) not technical enough (remembering that the vocal minority can have the most impact) but the happy ending is when the majority feel that the article/presentation was exactly what they needed.

It is often helpful for the would-be nontechnical writer (with a technical background) to look at how newspapers and magazines convey information—they use short paragraphs, strong headlines, sidebars and boxed out quotations. The writers also use eye-catching opening sentences and strong final paragraphs.

In fact, there are several failure modes that are evident when scientists and engineers write for a nontechnical audience:

- Too much 'how' and not enough 'why'—Reports written by scientists and engineers often contain page after page of the technical details of a particular solution that leave the reader confused.
- Too much detail—Completeness and accuracy are virtues that scientists and engineers have but often 100 well-chosen words will make the point better than 2,000 words covering every detail. When there is too much detail, the members of the audience lose patience and interest.
- Use of jargon and acronyms—Scientists and engineers find it easy to scatter acronyms, industry standard buzzwords, and shorthand terms, and assume that every nontechnical reader will know what they mean—making assumptions about what readers know is dangerous and foolhardy.
- Wanting to appear intelligent—Research shows that using short words makes writers seem more intelligent, but many people think they have to use long words and complicated sentences to appear smart. Too much scientific and engineering copy looks like the winner of a cuneiform writing competition translated into English.
- Use of humor—This is a dangerous tactic (that is generally employed during presentations rather than in written articles). The attempts at humor may receive a polite chuckle or be met by a stony silence—members of the audience might be offended by the attempts at humor.

As is the case in technical writing, audience analysis is a feature of all non-technical writing, but it may not need to be as extensive as for technical writing (Table 10.2). Nontechnical writing is a communication to convey a particular piece of information to a nontechnical audience for a particular purpose. It is often exposition about scientific subjects and technical subjects associated with scientists and engineers and their respective work projects and discoveries.

TABLE 10.2

General Comments on Nontechnical Writing*

Subject Treatment

The thoughts of writer may be included.

Scope of Document

Often depends on the interest and scope of knowledge of the writer.

Identification of Audience

Not always a specific audience and may be written for a general audience with varying knowledge skills.

Document Organization

May not be the same style of organization as a technical document but must have literary flow.

Concise

Must be concise and comprehensive.

---

* The comments for a technical document are generally diametrically opposite to those for a nontechnical document.

An important step in writing the article will be to carefully tailor the writing to the publication, which means that the scientist or engineer will need to fully understand the target audience (Chapter 4).

Procedural nontechnical writing translates complex technical concepts and instructions into a series of simple steps that enable readers to understand the nature of the work. To present appropriate information, nontechnical works must take the extra step to help the audience understand the goals.

Persuasive nontechnical writing attempts to sell products or change customer behavior by putting forth compelling descriptions of how a product or service can be used in one's life. This type of writing often delves into features and benefits of the product or service, and may use illustrations to make the benefits easier for the audience to understand.

The scientist and engineer must understand that he or she will have a very different level of technical background to the nontechnical person. It may even be a shock to realize just how vast this difference really is. The readers have not spent the thousands of hours addressing scientific and engineering problems that the writer has.

No matter how experienced a scientific or engineering writer may be, he or she always has to revise documents and tune them for the *real people* (the nontechnical audience). These are the people who will interface with a product in many different ways in many unexpected environments.

Thus, writing for a nontechnical audience involves: (1) analyzing the audience (Chapter 4), (2) selecting the right format (Chapter 2 and Chapter 3), and (3) carefully structuring the content (Chapter 3).

Overall, the writer must fight the natural tendency to attempt to impress the audience with his or her thorough knowledge of the subject. It is better for the writer to use his or her knowledge to decide what is most important for the audience to know and, by inference, to determine what is best omitted or delayed until the time of a future article.

By summarizing a topic in easy to read writing, a writer will allow the audience to pay attention to, and assimilate, the most important points of the article rather than confront the audience with a mass of unintelligible detail.

10.4 SCIENTIFIC AND ENGINEERING POSTER PRESENTATIONS

A scientific or engineering poster is a document that can be used to communicate the research finding of a scientist or engineer at a technical conference. It is essential that the design of the poster be such that it will be remembered rather than forgotten (Tufte, 1983; Woolsey, 1989; Block, 1996; Briscoe, 1996; Wolcott, 1997; Keegan and Bannister, 2003).

A poster for a presentation is typically composed of a short title, an introduction to the research problem question, an overview of the scientist's or engineer's experimental approach, the results, some insightful discussion of the results, a list of previously published articles that are important to the research, and acknowledgment of the assistance and financial support from others. If the text is kept to a minimum, a person could fully read the poster in less than ten minutes.

As is to be expected, the various sections of the poster or thesis must be clearly identifiable and appropriate, even to those who may be disadvantaged by color blindness (Rigden, 1999). The content of each section should correspond to the subtitle used and the transition from one section to next should be easy to follow. The terminology has to be uniform throughout the poster or thesis, any abbreviations should be consistent, and units of measurements should be the same in the text as in tables.

Above all, the writing style has to be clear and pleasant and there should not be spelling mistakes—special attention is needed in following the various British or American spellings (e.g., *centre* vs. *center*).

Although the author of the poster could communicate all of the above via a 15-to-20-minute presentation at the same meeting, presenting a poster allows the scientist or engineer to interact personally with the meeting attendees who are interested in the research, and also allows the author to reach people who might not be in his or her specific field of research.

Some scientists and engineers consider posters to be more efficient than a verbal/visual presentation because they can be viewed at any time during the meeting and are especially desirable if the author is not skilled in making presentations. In addition, once a poster has been designed and produced it can be taken to other conference venues (Purrington, 2009).

10.4.1 Layout

Unlike a manuscript, scientific and engineering posters can (and should!) adopt a variety of layouts depending on the form of graphs, tables, and photographs. The author should maintain sufficient white space, keep column alignments logical, and provide clear cues to the readers as to how they should read through the poster elements.

The poster may also be creative insofar as it does not follow the standard layout of a scientific or engineering manuscript paper (Chapter 3). If the author wishes to make some sections (that few people read) less prominent at, say, the bottom portion of the poster, this will make space available for the more important sections such as the *Conclusions*.

The best method of producing a first-class noticeable and readable poster is to get fully involved in process of producing a rough draft.

Rough drafts are especially crucial in deciding whether the scientist or engineer (and which scientist or engineer has not changed his or her mind at some time during the preparation of any type of presentation?) needs to delete text or add text or resize figures or change the fonts. Such decisions, while appearing trivial to some scientists and engineers, can make the difference between an interesting and readable poster and a ho-hum message that has little or no effect on the audience.

In terms of timing, a rough draft of the poster should be produced *at least* one month before the poster submittals date. At this stage, the author should wisely request that several colleagues examine it—these persons should be completely trustworthy and not give away the contents to others as part of a coffee-group conversation, or any conversation for that matter! The reviewers (for that is what they are) should be asked to comment on word count, prose style, idea flow, figure clarity, font size, spelling, figures, and tables.

If the author decides to print a miniature version of the poster on letter-sized paper to get a very rough sense of any layout changes that may be necessary, the shrunken version may be extremely hard to read and difficult to critique.

10.4.2 Parts of the Poster

As with any scientific or engineering document, the poster will be divided into several parts. It is during the planning of the poster that the author will have to decide which parts take prominence over others and the space to be allotted to each part.

The author of any poster will be well advised to understand that the typical attendee at a poster presentation may be seeking relief from listening to speakers drone on in monotones and needs to bring some color into his or her afternoon or evening. He or she may not even have any interest in the poster session and merely needs to stretch his or her legs. Thus the watchwords for any poster are *eye-catching*.

Nevertheless, a poster should contain several recognizable parts of which the most typical are the following:

- **Title**—The title should convey the nature of the problem being researched and the approach employed (maximum length: one to two lines); larger fonts are often used for the title to catch the eye of the passer-by, and the author would do well to remember that titles with colons often take longer to read (Lewison and Hartley, 2005) and may test the patience of the would-be viewer.
- **Abstract**—The need for an Abstract is variable and meeting-dependent. Typically, when a poster is presented at a meeting the author is asked to *submit* an abstract for inclusion in the meeting catalog or preprints. However, some organizations do require an Abstract to be included on the poster; if so, the Abstract should not be long—usually 50 words or less.
- **Introduction**—Once the interest of a potential poster-viewer has been caught by the Title, the Introduction is used to the get viewer(s) *more interested* (even euphoric) in the issue or question while using the absolute minimum of background information and definitions; the issue can be placed in the context of published, primary literature; or provide description and justification of the general experimental approach. Unlike a manuscript, the introduction of a *poster* is an ideal place for a photograph or illustration that communicates some aspect of the research question (maximum length: approximately 200 words).
- **Experimental Equipment and Methods**—The Experimental Equipment Materials and Methods section is used to describe briefly the experimental equipment and methods, but not with the detail used for a manuscript; figures and tables are ideal to illustrate experimental design and flow charts can be used to summarize reaction steps or timing of experimental procedures (maximum length: approximately 200 words).
- **Results**—The Results section is key to the whole poster and should include data analysis that more specifically addresses the hypothesis; opt for figures whenever possible and provide figure legends that could stand on their own

to convey important points to the viewer (maximum length: approximately 200 words, not counting figure legends).

- **Conclusions**—With the Results section, the Conclusions section is the crux of the poster; the viewer can be reminded of the hypothesis and result, and it should also be stated whether the hypothesis was supported. If the results do *not* support the hypothesis, the author should have questioned why he or she is at the meeting making a poster presentation; the relevance of the findings to other published work can also be stated (maximum length: approximately 200 words).
- **Literature Cited**—The manner in which the literature is cited should be the standard format required by the society hosting/sponsoring the meeting and/or poster session (maximum length: usually on the order of ten citations).
- **Acknowledgments**—It is in order to thank individuals for *specific* contributions to the project such as equipment donation, laboratory assistance, and comments on earlier versions of the poster, and the funding organization (maximum length: less than fifty words).

There are usually those viewers who wish to know more about the research, and the scientist or engineer can use this section to provide his or her e-mail address, a Web site address, and perhaps a URL where viewers can download a PDF version of the poster (maximum length: approximately twenty words).

10.4.3 OTHER ASPECTS OF THE POSTER

The major trap that scientists and engineers fall into is to make the poster too long. Densely packed, high word-count posters are basically manuscripts pasted onto a wall, and often scare away many would-be viewers. Posters with 800 words or less are ideal.

In addition to the author avoiding titles with colons, which are longer than a normal title (Lewison and Hartley, 2005) and so take longer to read (see above), the title should be formatted (usually in *sentence case*) so that the formatting does not obscure useful naming conventions that depend on font formatting. Usually a nonserif font (e.g., Helvetica) is preferred for title and headings and a serif font (e.g., Palatino) for the body of text (serif-style fonts are much easier to read at smaller font sizes).

It is preferable to avoid the use of dark backgrounds, which can make designing graphics extremely difficult. To make graphics work on a dark background, the author would need either (1) to invert the figures so that they stand out against a dark background or (2) to frame the figures in white boxes. Both of these are time consuming and the latter uses up white space unnecessarily. It is better to just use a light background.

Figures and tables are ideal for posters—the use of short, informative titles helps to *lead* the viewer more effortlessly through the poster. And, visual additions help attract and inform viewers much more effectively than text alone; tables are also a benefit. The author should make sure that details on graphs and photographs can be *comfort-*

ably viewed from six feet away. A common mistake is to assume that axes labels and numbers as well as figure legends are somehow exempt from font-size guidelines.

Incorporate *Web graphics* with extreme caution. Most Web images have 72 dots per inch of resolution, but *printing* at that resolution looks unprofessional and lacks clarity, and the figure will be a huge turnoff to prospective viewers. If a photograph is included in the poster, a thin gray or black border should be added to make it more visually appealing. Do not overpower the image with an overly thick line but choose a line color that is subtly pleasing but barely noticeable to the viewer.

REFERENCES

Block, S. 1996. The DOs and DON'Ts of Poster Presentation. *Biophysical Journal,* 71: 3527–3529.

Briscoe, M.H. 1996. *Preparing Scientific Illustrations: A Guide to Better Posters, Presentations, and Publications,* 2nd ed. Springer-Verlag, New York.

Hanson, K. 1999. *Writing for Trade Magazines.* Dixon-Price Publishing, Kingston, Washington.

Keegan, D.A., and Bannister, S.L. 2003. Effect of color coordination of attire with poster presentation on poster popularity. *Canadian Medical Association Journal,* 169: 1291–1292.

Lamott, A. 1994. *Some Instructions on Writing and Life.* Anchor Books, Peterborough, United Kingdom.

Lewison, G., and Hartley, J. 2005. What's in a Title? Numbers of Words and the Presence of Colons. *Scientometrics,* 63(2): 341–356.

Matthews, J.R., Bowen, J.M., and Matthews, R.W. 1996. *Successful Science Writing: A Step-by-Step Guide for the Biological and Medical Sciences.* Cambridge University Press, Cambridge, United Kingdom.

Purrington, C.B. 2009. Advice on designing scientific posters. http://www.swarthmore.edu/NatSci/cpurrin1/posteradvice.htm.

Rigden, C. 1999. The eye of the beholder—designing for color-blind users. *British Telecommunications Engineering,* 17: 2–6.

Tufte, E.R. 1983. *The Visual Display of Quantitative Information.* Graphics Press, Cheshire, Connecticut.

Wolcott, T.G. 1997. Mortal Sins in Poster Presentations Or, How to Give the Poster No One Remembers. *Newsletter of the Society for Integrative and Comparative Biology,* Fall: 10–11.

Woolsey, J. D. 1989. Combating Poster Fatigue: How to Use Visual Grammar and Analysis to Effect Better Visual Communications. *Trends in Neurosciences,* 12: 325–332.

11 Correspondence

11.1 INTRODUCTION

Written correspondence is essential to scientific and engineering practice. Formal and informal letters, e-mail letters, and memoranda (memos) allow the scientist or engineer to build and sustain relationships with colleagues (some of whom he or she may never meet).

Scientific or engineering correspondence (as used in the context of this book, which excluded correspondence in the form of brief communications for publication in the *Letters to the Editor* section of a journal) is a form of writing used by scientists and engineers to exchange information with their peers and may involve several degrees of formality.

In addition, because scientists and engineers are often busy people and focused on their own work, their written letters and e-mail should be concise and specific and they should use shorter paragraphs than in a formal scientific engineering paper or report. In addition, the writer must also carefully consider the audience (Chapter 4), which can also help the scientist or engineer write effective professional correspondence.

The exchange of scientific and engineering information is almost as old as writing itself and certainly was in operation once writing had been developed as a means of communication and not only for record keeping.

Writing emerged in many different cultures and in numerous locations throughout the ancient world. The Sumerians of ancient Mesopotamia are credited with inventing the earliest form of writing, which appeared circa 3500 BC. The writing was in the form of pictograms (using the end of a wedge-shaped reed, hence the term *cuneiform* writing) that were impressed into soft clay tablets (often on both sides of the tablet), which were then baked for hardening and preservation. It is no surprise that many of the clay tablets produced for record keeping and message transmittal were a convenient size. Many were the size of a modern cell phone, which allowed easy storage and carriage, usually in a pouch that was slung over the carrier's shoulder. This allowed news and ideas to be carried to distant places without having to rely on a messenger's memory.

However, for the present purposes it is more pertinent to trace the delivery of written material using the English Royal Mail system, which was probably the first public delivery system for written documents.

The Royal Mail traces its history back to 1516, when Henry VIII established a *Master of the Posts*, which eventually evolved into the office of the Postmaster General. The Royal Mail service was first made available to the public by Charles I on July 31, 1635, with postage being paid by the recipient. The monopoly was awarded to Thomas Witherings but in the 1640s Parliament removed the monopoly from Witherings and during the Civil War (1642–1651) and the Commonwealth (1649–1660, the English Interregnum) the parliamentary postal was run at a profit by Edmund Prideaux, a prominent parliamentarian and lawyer who rose to be Attorney General, although there are questions about the legality of his methods, even though he improved the efficiency of the service (How, 2003).

In 1653, the English Parliament set aside all previous grants for postal services, and contracts were let for the inland and foreign mails to John Manley, who was given a monopoly on the postal service, which was effectively enforced by the government of the Lord Protector Oliver Cromwell. Manley ran a much improved Post Office service. In July 1655 the Post Office was put under the direct government control of John Thurloe, a Secretary of State and Cromwell's spymaster general. Previous English governments had tied to prevent conspirators communicating, but Thurloe preferred to deliver their mail having surreptitiously read it.

At the restoration of the English monarchy in 1660, the General Post Office (GPO) was officially established by Charles II (Marshall, 2003). Between 1719 and 1763, Ralph Allen, Postmaster at Bath, signed a series of contracts with the Post Office to develop and expand Britain's postal network by organizing mail coaches for intercity transport of the mail (Smith, 2004).

In January 1840, a single rate for delivery anywhere in Great Britain and Ireland was prepaid by the sender. A few months later, to certify that postage had been paid on a letter, the sender could affix the first adhesive postage stamp, the Penny Black that was available for use from May 6 the same year. Other innovations were the introduction of prepaid postal stationery and envelopes. By the late nineteenth century, there were between six and twelve mail deliveries per day in London, permitting correspondents to exchange multiple letters within a single day.

For scientists and engineers, letters were the most common form of writing in the seventeenth century. Using the mid-1600s as a focal point, the period was dominated by the proliferation of scientific and engineering academies, printed journals, and correspondence networks.

The establishment of a system of mail delivery meant that letters were regarded as an effective, rapid, and certain means of communication and were, therefore, easily adapted to the needs of the emerging scientific and engineering disciplines as a means of exchanging information (Dawson and Gregory, 2009).

However humble and ordinary, handwritten letters were a powerful form of communication and a fact of daily life. By existing standards, letters were not only convenient and inexpensive but they were the fastest and most unrestricted form of science writing. From the first decades of the century, letters could be sent and received within weeks anywhere in Europe, and before the century was over, the "ordinary" post combined with diplomatic couriers to establish wider, faster, and more reliable service that extended to the Levant.

More than any other form of science writing, correspondence opens an historical window on these changes. Spontaneous and fresh, letters show science in the making. At the most practical level, letters help historians pinpoint the date of an observation or experiment, or more generally, they explain problem selection, changes in approach, or the fate of a failed hypothesis. Letters sometimes provide frank appraisals of the work of others, and often supply the only written record of private activities, friendships, and rivalries, not to mention the collaborations, controversies, and inner workings of informal groups. Unlike the printed book, letters repeatedly fail to separate public and private.

From the earliest decades of the Scientific Revolution, letters foreshadowed and underwrote subsequent forms of *scientific exchange*. Handwritten letters were daily facts of life that entered a scientist's or engineer's work. Written correspondence was essential to scientific and engineering practice. Letters, e-mail, and memos allow the writer to build and sustain relationships with his or her colleagues, so everything that is written should represent the writer's character and abilities fairly. Because scientists are often busy people, written letters and e-mail should be concise and specific. Readers tend to look at letters, e-mail, and memos quickly, so shorter paragraphs than in a formal scientific paper or report should be used.

In addition, and by analogy to other written works, the writer needs to consider the audience and purpose of the communication, to whom he or she is writing, what is hoped to be accomplished, and the tone of the correspondence. Thinking carefully about the audience and tone can help a scientist or engineer answer these types of considerations and write effective professional correspondence.

11.2   TYPES OF CORRESPONDENCE

Generally, for the scientist and engineer (and other professionals) there are two types of correspondence (1) business correspondence and (2) social correspondence. It is not the purpose of this text to get involved in descriptions of compiling and the use of social correspondence. In fact, for the purposes of this text, business correspondence is further subdivided into memoranda (memos), letters, and electronic mail, which includes fax correspondence and e-mail correspondence that might be sent from one scientist or engineer to another. The more formal types of (nontechnical) business letters are described adequately elsewhere (Poster and Mitchell, 2007).

In engineering and science, correspondence is an effective way to make requests, submit changes to a job, and deliver specific information. Unlike telephone conversations, correspondence presents the audience with a legal contract that is dated and can support a claim in court. This section presents formats for memos and letters. Because electronic mail usually has a built-in format, no format is assigned here for it.

Not one modern communications marvel can replace a letter. A letter can have special powers: It can be more intimate and touching than even a conversation. It can be more personal than any telephone call.

Many scientists and engineers have trouble getting started but, once started, can continue comfortably. It is a good idea to mentally go over the main things before starting. As a cardinal rule, it is often best to resist opening a letter by apologizing for not writing sooner.

In correspondence, a scientist or engineer should concentrate on being clear and precise. Because audiences tend to read letters and memos quickly, a writer should opt for shorter sentences and paragraphs than would be used in a formal report or journal article. Also, the tone of the correspondence should receive very careful consideration.

Tone is difficult to control in most types of correspondence. Often, scientists and engineers lose control of tone by avoiding simple straightforward wording and when scientists or engineers write a business letter or memo, they change their entire personality. Instead of using plain English, there is a tendency to use *buzzword* phrases such as *per your request* or *enclosed please find*. These phrases are not natural or straightforward and they tend to inject an undesirable attitude—which may be perceived as an arrogant attitude—into the writing.

Thus, the general forms of writing for scientists and engineers include (1) letters, (2) memoranda, (3) faxes, (4) e-mail communications, and (5) progress reports. The necessary constituents of a progress report are included here as well as elsewhere (Chapter 2) because the progress report can take any one of several forms. The report may appear as (1) a letter, (2) a memorandum, (3) a fax, or (4) an e-mail, depending on the request of the funding agency or the project sponsor. In addition, the format of the progress report will be made clear in the contract or, if not, by the funding agency or the sponsor.

Job search correspondence and résumé writing, although important to many scientists and engineers, is not included here. However the importance of such correspondence in a job search should not be underestimated as it is crucial to the success of the job seeker. For works related to job search correspondence and résumé writing, the reader is referred to online advertisements for courses and books (see, for example: Frank, 1993; Rosenberg, 2008; Vick et al., 2008; Bly and Kelly, 2009).

11.2.1 LETTERS

The letter was originally designed as a means of communication for a small, diverse, and isolated community (Goodman, 1994) and was particularly suited to conditions where the writer and reader rarely had the chance to meet. The speed of the transmittal of the letter made it useful for organizing laboratory or field events, which are a necessary part of science and engineering, and for disseminating and comparing information from widely dispersed sites.

Generally, scientific and engineering letters from one scientist or engineer to another have been relatively free from censorship. Indeed, various forms of correspondence networks were plentiful throughout the nineteenth century and offered a means of freedom of expression through letters (Dawson and Gregory, 2009). However, in some cases, committing information to letters could be dangerous; but with proper precautions, controversial issues circulated freely under the cloak of secrecy or under the cloak of friendship. By such means, questionable opinions (usually political opinions rather than opinions pertaining to science or engineering) could be expressed with a meaning clear to the recipient but not (should the letter be intercepted) to a potential censor.

From this early form of communication, there evolved the use of business letters, which became the main way businesses officially communicate with their customers and other businesses. To a lot of people business letters are still synonymous with business correspondence, though nowadays the phrase business correspondence tends to have a much broader meaning and may, in a court of law, include any form of transmittal (paper or electronic) from one person (or several persons) to another person (or persons) or from one person (or several persons) *to file*.

As with any other letters, business letters are the product of the person (or persons whose signatures are attached) writing them and depend on the personal preferences of the writer(s). On the other hand, there are rules for writing business letters, some of which are strict while others can be ignored—sometimes breaking some of the rules (those that are obsolete and others that need to be broken) can even make a letter more effective.

In the modern business world, letters remain a staunch form of communication and formats for letters vary from company to company and even within government departments.

People read business letters quickly. Therefore, get to the point in the first paragraph—the first sentence, if possible. In other words, the writer should state what he or she wants up front. Shorter sentences and paragraphs should be used than would be used in a longer document—sentences are generally held to less than twenty words, and paragraphs held to less than seven lines.

The letter should be spaced on the page so that it does not crowd the top or the bottom of the page. In addition, it is also preferable that a letter be limited to one page—the second page of a letter is not often read. As with memos, anyone whose name is mentioned in the letter or who would be directly affected by the letter should receive a copy of the letter.

Formal letters differ in some key ways from e-mail and even from professional memos. The purpose of a formal scientific and engineering letter is often serious. Such letters contain information that is too important to handle in an informal medium like e-mail or a memo. In addition, recipients may want a document that can be signed and filed in a traditional way.

Formal letters should end with *Sincerely* and progress toward familiarity with (1) *Yours truly*, (2) *Regards*, (3) *Best wishes*, or similar appropriate wording.

The final words for any such communication *before it is sent to the recipients* are (1) check, (2) check again, and (3) check one more time.

11.2.2 MEMORANDA

A memo (the shortened version of *memoranda*) is a document or other communication that helps the memory by recording events or observations on a topic, such as may be used between scientists and/or engineers.

Like e-mail messages, memos are common in many workplaces. A memo may serve as an informal proposal to propose a new idea to a supervisor or manager. It can also provide a quick, concise way for scientists to brief each other or their supervisors about the status of a project.

Memos are less formal than scientific papers or lengthy technical reports, but they should still show a respectful and professional tone. Unlike e-mail messages, memos should remain formal even if the audience is well known to the writer.

As with any writing task, the scientist or engineer needs to consider the reason for the memo, the potential readership, and the focused interests of the readers in the context of the memo. Considering these issues will help the writer determine the appropriate tone and structure for the memo.

In many instances, a memorandum is not written to inform the reader but protect the writer, especially when a scientist or engineer has a potentially patentable idea that will become company intellectual property leading to promotion or financial reward. Such a memorandum might, within a company, be called a *patent memorandum*, which should be signed and dated; it could be produced in court as evidence to establish *rights of ownership.*

As with e-mail, the writer should consider very carefully the recipient of the message and what the writer hopes to convey before preparing the memo. For example, if a scientist or engineer is writing a memo to propose a new project to his or her supervisor, he or she must explain why the project is necessary and worthwhile. If the scientist or engineer is updating the reader on the status of a project, he or she may need to focus on how much the project has cost so far and when it will be completed—under or over budget.

Typically, scientists and engineers write memos to people within the workplace, and letters are written to people outside the workplace. One major difference between memos and letters is the title line found in memos. Because readers often decide whether to read the memo solely on the basis of this title line, the line is important. Another difference between letters and memos is that memos are sometimes written to serve as short reports. In such cases, the format for the memo must be changed. For instance, in a memo serving as a progress report for a project, the writer might include subheadings and sub-subheadings.

In memos that make requests or announcements, the writer should keep the sentence lengths and paragraph lengths relatively short. Sentences should average fewer than twenty words, and paragraphs should average fewer than seven lines. Also, the writer should attempt to contain the total memo length to less than one page whenever possible.

Sometimes companies use memos to communicate short reports (two pages or more). For these types of memos, the format will be changed from the typical memo. For instance, illustrations can be included, appendices attached, and the text can be subdivided into sections. If references are necessary, a list should be included at the end. In memos that act as reports, the style changes as well. For instance, the sentences and paragraphs are typically longer than in memos that simply provide announcements or make requests.

The goal of a memo is to convey essential information quickly, so the writer should not distract the audience even if he or she (the writer) is trying to be friendly. For this reason, memos typically do not include greetings or closings. Choosing the tone carefully is especially important if the writer needs to deliver bad news in the memo.

For example, if a scientist or engineer is updating his or her manager to tell him/her that his or her project is running behind schedule, the memo should be forthright and honest. There should be no sign of false cheerfulness or optimism. It is the professional responsibility of every scientist and engineer to explain the situation exactly as it is, not to withhold bad news to keep the audience happy. When there is bad news to deliver, words like *unfortunately* or phrases like *I regret to tell you that* can help express the writer's dismay, but there should always be an explanation of how the problem can be solved.

While the format for a memo may not be specified, many organizations and institutions do have guidelines that specify the format. The writer should adhere to such guidelines. In fact, some upper-level managers have been known to refuse to read any document that is not formatted according to the guidelines specified by the organization.

Thus, if an organization has a set format for memos, the scientist or engineer must follow that format. Both e-mail and memos feature certain information in their headers, but unlike e-mail, memos do not include a salutation or a closing. As with e-mail, the body of a memo may include headings, subheadings, or bullet points to highlight important information—recognizing that too many bullet points will make the most important ideas difficult to identify.

A memo may be one page long or many pages in length. If the user is a senior executive manager or chief executive officer, the format might be rigidly defined and limited to one or two pages. If the user is a scientific or engineering colleague, the format is usually much more flexible. At its most basic level, a memorandum can be a handwritten note to one's colleague or supervisor.

Generally, the memo should be spaced on the page so that it does not crowd the top. Also, copies should be sent to anyone whose name is mentioned in the memo or who would be directly affected by the memo. Finally, the final paragraphs of memos can be used to make requests or announcements to inform the readers what the writer needs them to do or what he or she will do for them.

If the writer mentions colleagues in a memo, they should receive a copy of the memo with their names listed on the *cc* line. In addition, if the writer needs to include another document (such as a preliminary budget or a detailed timeline) as an attachment, this should be noted in the memo and the title of that document included.

The final words for any such communication *before it is sent to the recipients* are (1) check, (2) check again, and (3) check one more time.

11.2.3 FAXES

A fax document (short for *facsimile document*) is a document that is sent over a telephone line. For decades, and even at the time of writing this book, using a fax machine (facsimile machine) has been considered to be the easiest way of sending paper documents. In fact, the use of fax machines for business communication became common in the 1970s and 1980s but for a long time many scientists and engineers did not have access to fax machines.

In more general terms, faxes have been a recognizable part of the business environment for approximately forty years, which is a short period compared to the longevity of the use of business letters. Consequently, there are not as many rules established for writing faxes. They were written in the way that was considered appropriate and have still retained a personal touch, with the ability of scientists and engineers to send faxes via computer, which is in that sense similar to e-mail.

There are still many countries where the Internet and newest computer technologies are not as common as they are believed to be. In fact, even in the developed countries some individuals would not come close to a computer or would just use some very basic computer features and think that the rest is too complicated to learn. Hence, in many offices a fax machine can still be one of the most common types of machines for transmittal of business correspondence.

Faxed messages are still a viable means of correspondence for scientists and engineers. Many prefer to send by fax documents that have not originated on a computer or have not been scanned into a computer to a colleague for comment.

Fax machines still retain some advantages, particularly in the transmission of sensitive material which, if sent over the Internet unencrypted, may be vulnerable to interception. In some countries, electronic signatures on contracts are not recognized by law while a faxed contract with a signature at the bottom of a document will be accepted as a legal document.

Furthermore, handwritten forms are still in use and the fax machine is the best option for sending such items.

The essence of the fax transmittal is the cover sheet—it is a close (but not exact) analog of the envelope for the letter, although there are many instances where faxes are sent without cover sheets. Many companies have specially designed fax cover sheets which contain information that (1) identifies the recipient, (2) helps the recipient identify the sender, and (3) distinguishes a fax message from other faxes that may be coming through at the same time. A disadvantage is when a large office still uses one fax machine for the whole office and delivery of the incoming fax to the intended recipient in a timely manner is not always assured.

Dealing with a pile of faxes that fell from the fax machine onto the floor is a daunting task and in such instances cover sheets are truly appreciated, especially when they note the number of pages that are supposed to be in the fax. Usually, the cover sheet is included in the page count, although modern faxes usually (but not always!) have the page count printed on the transmitted pages.

This has been circumvented to a great extent by the direct delivery of a fax message to a recipient's computer.

Nevertheless, fax cover sheets must at least include the fax number of the sender and the recipient as well as the number of pages—the last inclusion make sure that the recipient is aware of any pages that might be missing through faulty transmission or have been picked up by another recipient of a different fax in the same office; such pages are often discarded without thinking by the nonrecipient.

The use of a cover sheets is the decision of the sender. However, if he or she is sending a fax to a company the sender has never dealt with, the situation certainly calls for a cover sheet. If the fax is being sent to a colleague who always uses a fax cover sheet and has standard procedures for fax transmissions, it's preferable to do the same. On the other hand, if the sender repeatedly sends faxes to the same person in an established business relationship, the need for a cover sheet may be diminished.

There is no specific design or pattern to a fax cover sheet, but it is preferable if the information is organized in a manner that doesn't confuse the fax recipient. Many companies opt for a fax cover with a heading similar to that of a memorandum. Company letterhead is practically always used nowadays for the cover sheet heading. Many companies include the Web site, which might be a good idea for the company, if it is not already being done.

The final words for any such communication *before it is sent to the recipients* are (1) check, (2) check again, and (3) check one more time.

11.2.4   E-MAIL

Electronic mail (*e-mail* or *email*) is a method of exchanging digital messages from an author to one or more recipients.

E-mail is helpful as a means of instant communication but may be a burden for those scientists and engineers who do not rigorously check their e-mail Inbox several times each day. Spam (*junk email, unsolicited bulk e-mail*) is also a very big issue though a little less so lately when there are ways to harness it (sometimes but not always successfully). Several governments have passed anti-spam legislation.

E-mail technology operates across the Internet through which e-mail accepts, forwards, delivers, and stores messages. The principal advantages of electronic mail over other types of correspondence are its speed and ease of use. For instance, in minutes, information can be sent to many recipients around the world. Neither the users nor their computers are required to be online simultaneously; they need connect only briefly, typically to an e-mail server for as long as it takes to send or receive messages.

For scientists and engineers, an e-mail message is a less formal version of a memorandum or a letter.

Scientists and engineers use e-mail communications to make requests, to answer questions, and to give announcements. When a scientist or engineer is communicating with someone professionally, however, e-mail is more important. In fact, e-mail is often the main mode of communication for scientists and the manner in which an e-mail is written can shape what other scientists think of the character of the writer.

As with any document, a well-written e-mail can impress the reader and show that the writer is thoughtful and responsible, whereas a poorly written e-mail can damage productive relationships or keep the writer from forming new ones. Therefore, before sending an e-mail, the writer must carefully consider the audience and the tone to be used.

E-mail communication can be read quickly—even from personal cell phones if the system is set up to receive and transmit e-mail messages. For that reason, the writer should get to the point in the first sentence, if possible. In other words, the writer should state what is required up front. On the other hand, complaints are usually better handled in person rather than in the form of a disembodied e-mail.

In an e-mail, sentence lengths and paragraph lengths should be kept relatively short and single-spaced. Sentences should average fewer than twenty words, and paragraphs should average fewer than seven lines. A line can be skipped between paragraphs, and a typeface that is easily read on a computer is necessary—many personal preferences for typeface include the use of Times New Roman. If possible, the total e-mail length should be kept to a length that can be viewed entirely on the screen.

Because the reader sees only the title of an e-mail in the Inbox or in the folder where it has been filed, the writer should give considerable thought to the title. The title should orient the reader/recipient to the subject of the e-mail and, if possible, distinguish the e-mail from other e-mails about that subject. For example, a title such as *A Method for Increased Distillate Yield from a Delayed Coking Unit*, while seemingly verbose, would quickly orient the reader/recipient to the subject matter.

As with other correspondence, copies should be sent to anyone whose name is mentioned in the e-mail or who would be directly affected by the e-mail. It is discourteous for such persons to hear that their names were used or thoughts quoted from a third party. Also, the writer should be sure to mention any attachments. Finally, the final paragraphs of e-mail correspondence should generally inform the recipients what the writer required them to do (if anything) as well as what the writer will do for them (if anything). Even a simple statement such *no action required on your part* is appreciated by many recipients rather than be left in the dark and have to make enquiring telephone calls.

One disadvantage of electronic mail is the crudeness of the format. Many electronic mail systems still do not allow the use of tabs, italics, and other formatting necessities that can be used in typical computing work. For that reason, the look of the message may not be as attractive as a memo or letter that has been printed on letterhead paper. Because the message does not look formal, many writers may mistakenly adopt a style that lacks the appropriate formality (Markel, 1996) and is more appropriate for use in a personal text message, as is commonly employed by many cell phone owners. . For instance, these people include needless abbreviations (such as *btw* rather than *by the way*).

Another disadvantage of electronic mail is also the ease of use. With letters and memos, the writer must print out the correspondence before it is sent. That printing task allows the writer to view the writing on paper—a step that makes it easier for him/her to proof for mechanical mistakes in spelling, usage, and punctuation.

With electronic mail, the writer does not have to print the message before sending it to the recipient(s). For that reason, electronic messages often are not as well proofread as is regular correspondence. The writer is advised to take the same care with electronic mail as he or she would with printed correspondence—the appropriate formality in style should be used as well as carefully proofing the message before it is sent.

In summary, e-mail correspondence to a colleague or peer must be (1) balanced, (2) respectful, (3) have an appropriate salutation, (4) contain enough information to show the purpose of the e-mail, and (5) a closing thanks for the recipient's time. This type of message will capture the recipient's attention and, most likely, the writer will receive a meaningful response.

The final words for any such communication *before it is sent to the recipients* are (1) check, (2) check again, and (3) check one more time.

11.2.5 Progress Reports

A progress report is written to inform a supervisor, associate, or customer about progress that has been made on a project over a certain period of time. The project can, in the current context, be any scientific or engineering study or research of a problem or question, or the gathering of information on a technical subject.

The necessary constituents of a progress report are included here rather than elsewhere (Chapter 2) because the progress report can take any one of several forms. The report may appear as (1) a letter, (2) a memorandum, (3) a fax, or (4) an e-mail, depending on the request of the funding agency or the project sponsor. The format of the progress report will be made clear in the contract or, if not, by the funding agency or the sponsor.

The progress report should summarize recent and future work on a specific scientific or engineering project. While the exact content and format of a progress report may vary from organization to organization, the purpose is universal and is to inform the recipient(s) if the work is (1) proceeding according to plan, (2) if any problems have been encountered, and (3) whether the original project plan is still viable, which should also include comments on whether or not the project can be completed on time and within budget.

Typically, scientists and engineers write progress reports when it takes well over three or four months to complete a project. A progress report may be in a short form (letter, memo, fax, or e-mail) or in a long form (technical paper) (Chapter 2 and Chapter 3) in which recent and future work on a specific project is summarized.

Thus, the same basic rules also apply to progress reports—which are a specialized type of memo. Managers usually review written progress reports quickly, so emphasize key ideas and important issues at the top of any paragraphs or bulleted lists. State early in the memo whether the work will be completed on time and on budget, and also note any problems that have been encountered.

In the progress report, the scientist or engineer (the author) should explain any or all of the following: (1) the extent of the completion of the work, (2) the work in progress, (3) the work that remains to be done, (4) problems or unexpected issues that have arisen, and (5) general comments on the progress of the project.

Thus, a progress report has several important functions: the report reassures the recipients (the funding source and internal managers) that (1) progress is being made, (2) the project is proceeding according to plan, and (3) the project will be completed by the expected date.

In addition, the report provides the recipients with a brief look at some of the findings or some of the work of the project. It also gives the recipients a chance to evaluate the work on the project and to request changes. Moreover, the report gives the author(s) the opportunity to discuss and evaluate any problems that have arisen and (most important) to forewarn recipients (under the auspices of a *no-surprises-at-the-end-of-the-project* courtesy). It also allows the author(s) to establish a work schedule so that they'll complete the project on time.

Furthermore, the recipient(s) of a progress report want to see whether the scientist(s) or engineer(s) involved in the project accomplished the goals specified in the original proposal (Chapter 2), what they are working on at the time of writing the report, what they plan to work on next, and how the project is going in general. To report this information, a progress report usually summarizes work within each of the following: (1) work accomplished in the preceding period, (2) work currently being performed, and (3) work planned for the next period.

The progress report may, subject to the formatting specified by the clients (recipients), also need (1) an introduction that reviews the history of the project's beginnings as well as the purpose and scope of the work, (2) a detailed description of the project, and (3) an overall appraisal of the project to date, which usually acts as the conclusion.

The introduction will allow the recipients to review the details of the purpose, scope, and activities of the project. This will aid recipients who are unfamiliar with the project, who do not remember certain details, or who want to double-check the approach to the project. The introduction can contain the following: (1) purpose of the project, (2) specific objectives of the project, (3) scope, or limits, of the project, (4) date the project began and date the project is scheduled to be completed, (5) personnel working on the project, and (6) overview of the contents of the progress report.

Not surprisingly, the most important part of the progress report is the Introduction. Here, the audience may need a reminder what the project is and why it is important. There should be an explanation of who is affected by the project, when the work began, and when it is expected to end. Finally, the overall status of the project should be outlined in specific terms so readers can see at a glance the status of the project and what remains to be done.

The body of a progress report should open by noting the current status of the project and state (1) any important tasks that have been completed, (2) any decisions that have been made, (3) any discoveries that have been made, and (4) the work that needs to be completed. Chronological order should be employed to show the audience what steps are yet to come and how long those steps may take.

Even though it seems counterintuitive, the writer should also describe any problems that have arisen during the project—the audience needs to know if something went wrong along the way, and they will want to know how the writer has responded. If those problems have been solved, the methods used should also be explained to the audience. If the problems were not solved, the writer should show that there is at least one possible solution to be tried. There should also be indications that the writer is considering other problems that might arise. This will show the audience that the writer has thought carefully about the project and how it will be completed.

The progress report should end with a summary of the current status of the project as well as any key issues. There should be a repeat statement whether the project will be completed on time and on budget.

Finally, there is the need to thoroughly check and revise the report before it is sent to the recipients.

As the scientist or engineer reads and revises the progress report, it is necessary to pay attention to items such as the following:

- *Format.* Make sure the correct format was used. The memo format is for internal progress reports; the business-letter format is for progress reports written from one external organization to another. Whether the author uses a cover memo or cover letter is often optional or dependent upon the request of the funding organization.
- *Introduction.* State that the document is a progress report, and provide an overview of the contents.
- *Headings.* Headings should be used to differentiate between the various parts of the report, particularly the different parts of the summary of work performed on the project.
- *Contents.* The author(s) should provide specifics and not make general statements about the work that has been performed. There should also be statements about the timing of the work (insertion of a Gantt chart is helpful

to show the progress of the project compared to the timeline presented in the original proposal).

- *Audience.* The author may also be sending the report to one or more nonspecialists. This should not cause the author(s) to omit discussion of technical aspects of the project—they can be written at a level that nonspecialists can understand or inclusion of a suitable nonspecialist (nontechnical) abstract will help the nonspecialist members of the audience.

The final words for any such report *before it is sent to the recipients* are (1) check, (2) check again, and (3) check one more time.

11.3 SUMMARY

Communications in any of the forms described above between scientists and/or engineers remain a formal way of communicating. In fact, the use of electronic letter (e-mail) is increasing. Although there are many different uses of such letters, typically they are informational.

As always, the most important aspect of writing any communication is the ability of the scientist or engineer to identify and write to the audience (Chapter 4). If the writer is addressing the communication to a less professionally skilled friend or colleague, he or she should avoid using highly technical terms that only scientists and engineers (with equal or greater skills than the writer) would understand. In addition, the writer should make sure that he or she presents the objective of the communication in a clear and concise manner. Another important element is for the writer to remain professional and to remain courteous.

Advances in electronic technology have dramatically changed the way in which scientists and engineers collect and use technical information. Indeed, such information can be transferred globally and more rapidly now than ever before and this has influenced the way many scientists and engineers think about privacy and the protection of personal information.

In fact, it is worth remembering that (at the time of writing) electronic forms of communication in many countries are less secure and do not have the same protection as conventional forms of communication, such as mail delivered through a national postal service. Thus, the same degree of confidentiality cannot be ensured for e-mail correspondence as for other forms of communication.

While the writer/sender of an e-mail has a reasonable expectation of privacy and that such communications should remain confidential and privileged, messages can be read and modified in transit, even before they reach their destination.

The moral of the story here related to the use of electronic forms of communication, especially e-mail, is very simple: *writer/sender beware*!

REFERENCES

Bly, R.W., and Kelly, R.A. 2009. *The Encyclopedia of Business Letters, Faxes, and E-mail.* Career Press, Pompton Plains, New Jersey.

Dawson, N.V., and Gregory, F. 2009. Correspondence and Coherence in Science: A Brief Historical Perspective. *Judgment and Decision Making*, 4(2): 126–133.

Frank, W.S. 1993. *200 Letters for Job Hunters*. Ten Speed Press, Berkeley, California.

Goodman, D. 1994. *The Republic of Letters: A Cultural History of the French Enlightenment*. Cornell University Press, Ithaca, New York.

How, J.S. 2003. *Epistolary Spaces: English Letter-Writing from the Foundation of the Post Office to Richardson's Clarissa*. Ashgate Publishing, Farnham, Surrey, United Kingdom.

Markel, M.H. 1996. *Technical Communication: Situations and Strategies*. St. Martin's Press, New York.

Marshall, A. 2003. *Intelligence and Espionage in the Reign of Charles II, 1660–1685*. Cambridge University Press, Cambridge, United Kingdom. P. 79.

Poster, C., and Mitchell, L.C. (Editors). 2007. *Letter-Writing Manuals and Instruction from Antiquity to the Present*. University of South Carolina Press, Columbia, South Carolina.

Rosenberg, A.D. 2008. *The Résumé Handbook: How to Write Outstanding Résumés & Cover Letters for Every Situation*. Adams Media, F+W Media, Avon, Massachusetts.

Smith, D.J. 2004. *Discovering Horse-Drawn Vehicles*. Shire Publications, Oxford, United Kingdom. P. 52.

Vick, J.M., Furlong, J.S., and Heiberger, M.M. 2008. *The Academic Job Search Handbook*. 4th edition. University of Pennsylvania Press, Philadelphia, Pennsylvania.

Glossary

Abbreviation: A word or phrase that is shortened, usually to a single recognized word format; the shortened form of a word or phrase, identified as an abbreviation by a dot/a full stop at the end of this format; a dot is usually only used when the abbreviation is a straight shortening rather than a common rewriting or omission.

Abstract: A summary in the beginning of a formal report or proposal; a synopsis of a scientific paper that often summarizes each of the paper's sections; written for a scientific or engineering purpose; sometimes incorrectly called an Executive Summary; see Executive Summary.

Academic freedom: The liberty or privilege that academics enjoy in regard to teaching, research, and publications.

Academic style: Rules for preparation of documents that were developed by well-known academic bodies around the world (some for specific scientific purposes and some for much broader academic objectives) and adopted by academic institutions and scientific and engineering authors as appropriate to their work.

Acronym: A word (usually pronounceable) formed from the initial letters of a name; used for legal and political acronyms in common parlance in some scientific and engineering disciplines.

Addendum: A separate document added at the end of the key document; in a published work, an addendum is usually referred to in the main text and added at the end of the main document to add meaning or explanation or to reduce the need for the reader to conduct additional research.

Advance: A percentage of the money paid to the writer by a publisher prior to publication of the book—advances are paid against future royalties and are paid back to the publisher once the book starts earning royalties. Writers of scientific and engineering book very rarely receive an advance royalty payment.

Advisers: Scientists or engineers who gather detailed information and provide information to decision makers.

Algorithm: The logical sequence of operations to be performed in the execution of a program.

Allegory: A narrative technique in which the characters are portrayed as things or concepts in order to convey a message; usually used for satirical or political purposes and not for scientific or engineering purposes.

All rights: The publisher owns all the rights to the work in all the media worldwide, but may not always own the copyright.

Alphabet-number system: A system of literature citation in which references are arranged alphabetically in the *Literature Cited* section but by number in the text.

Ambiguity: Allows for two or more simultaneous interpretations of a word, phrase, action, or situation, all of which can be supported by the context of a work; can cause confusion in scientific and engineering writing.

Ampersand: The "&" sign; usually used only in company name or joint venture formation; sometimes used in report headings (but not in the text); often discouraged in modern styles.

Analogy (*n* analog): A comparison of two unlike things, used to explain or illustrate a concept.

Appendix: Text that is near the end of a book. Usually, an appendix contains (1) information that helps some readers but which others do not need and (2) information that is not generally important but which readers will wish to know.

Applied research: The investigation of phenomena to discover whether their properties are appropriate to a particular need or want, usually a human need or want. In contrast, basic research investigates phenomena without reference to particular needs and wants. Applied research is more closely associated with technology, engineering, invention, and development. Basic research is sometimes described as *pure research.*

Article: A scientific or engineering paper which is a part of a journal that has a boundary separating it from the rest of the publication, that can be read and understood on its own, and whose author is identified; often a brief essay or report of known or likely interest to the audience of a broader publication, such as a scientific or engineering trade magazine.

Assignment: A piece of writing that a scientific or engineering writer has been assigned to write by an editor or publisher for a predetermined fee; in the life of a student this could be a term paper for which a fee is *not* paid.

Audience: The persons intended to be reached or to receive or to take an interest in the document; the readers of the document who need to understand all aspects of the document; a particular document may have more than one audience and may need to find a way of addressing the needs of more than one audience.

Audience analysis: Identification of the requirements or field of expertise of the audience of the members.

Author: The composer of a document, as distinguished from an editor, reviewer, or translator.

Author's guidelines: A set of guidelines that a publication wants the authors to follow.

Back matter: Features of a communication that appear after the last chapter or section such as appendixes, glossaries, and indexes.

Basic research: Investigation of phenomena without reference to particular needs and wants; sometimes described as pure research.

Bibliography: A list of books, magazines, journals, people, websites, or any other resources that were consulted in the process of writing a book, article, or paper; usually, the documents in the list are related to the content of the primary document and contain background information.

Bimonthly: Once every two months.

Bionote (Bio): A short two- or three-sentence description of the author written in the third person, usually to accompany an article.

Biweekly: Once every two weeks.

Boldface: Text in bold; often boldface is used to imply reference or meaning particular to this document and, since there are no agreed uses of boldface, this usually needs to be explained somewhere, often in the glossary; or maybe in a footnote: the footnote may be a footnote to the Preface or a footnote to the first bolding to appear in the document; the footnote will usually be the preferred method where the boldface is used to imply that the word appears in the glossary; the Preface is preferred in a technical report because there is no guarantee that the document is going to be read in a linear fashion. Because of this implied particularity, some forms of bolding should be removed in mark-up (such as emphasis) to be replaced by more agreed forms (such as italicized face for emphasis) that can be more easily understood without explanation.

Bolding: See Boldface.

Brainstorming: Generating ideas as quickly as possible, withholding evaluation of those ideas until later.

Business writing: Any kind of writing assignment done specifically for a business; can be subdivided into (1) writing internal documents for circulation within the business itself and (2) writing for clients and customers outside of the business.

Buzzword: A word used to create interest or excitement in the text or generate other text in response.

Canon: Works generally considered by scholars, critics, and teachers to be the most important to read and study; more frequently applied to papers or laws relating to religious doctrines.

Caption: A brief description of a picture, graph, table, or diagram; a label for a graphic (figure) or table; the explanatory comment accompanying an illustration; the heading of an article or document.

Cause and effect: A way to organize a communication that helps readers understand the relationship between one topic caused by another.

Citation: A formal entry directing the reader to previous work that serves as proof or support; see Reference.

Classification: Arranging information into groups that are related.

Cluster sketch: Writing the overall topic in the center of a page and then adding subtopics, joining them to the main topic or each other.

Code of conduct or Code of ethics: A central guide and reference for users in support of day-to-day decision making. It is meant to clarify an organization's mission, values, and principles, linking them with standards of professional conduct. As a reference, it can be used to locate relevant documents, services, and other resources related to ethics within the organization.

Complex audience: The diverse group of people who may read a written work from different perspectives.

Conflict of interest: A person has a conflict of interest when the person is in a position of trust that requires him or her to exercise judgment on behalf of others (people, institutions, etc.) and also has interests or obligations of the sort that might interfere with the exercise of this judgment, and that the person is morally required to either avoid or openly acknowledge.

Connotation: Implications that go beyond the literal meaning of the word; often frowned upon in science and engineering because of the potential for misinterpretation and ambiguity.

Content writing: Providing relevant information or text content specifically for Web sites; the writer must be skilled at research, analysis, and able to adapt to meet the needs of their clients.

Contexts of research: Laboratory versus field studies, experimental versus observational designs, individual-investigator versus collaborative projects, high versus low stakes outcomes, studies that are local versus national versus international in scope, insufficient versus adequate resources to achieve the goals of the research.

Copy edit: Prepare a document for presentation in a printed form; typically the term copy edit is used to describe the kind of light editing in which errors of style, grammar, word usage, and punctuation are corrected; checking for errors in spelling, grammar, punctuation, and word usage.

Copyright: The ownership by an author of his or her work. Copyright laws recognize the author's right of ownership of anything that the author writes immediately upon its creation; the legal right to publish a document—an author automatically has the copyright of a document, but the author can assign copyright.

Copywriting: The art of creating content (or *copy*). Usually, the term refers to writing in the literary sense, instead of the language that scientists and engineers create. For example, copywriters create text for journals, magazines, brochures, and other types of marketing communications.

Copyright law: Set of laws that determine whether an author can use other people's writing without their permission.

Cover letter: In the current content, a letter written as an introduction to a research paper, which gives a brief overview of what is described in the paper; the cover letter should not be an Abstract but should sum up some key points of the article and the author should present arguments as to why the article is worth publishing—with the advent of online submittal systems, the cover letter is becoming extinct.

A letter written to an employer, which briefly states the reasons why an applicant should be considered for a job position; a cover letter should be specific to the advertised job and should not be form letters.

Cross-reference: A direction from one part of a document either to another part of the document, or to a different document. In a printed document, a cross-reference usually contains a page number. In an online document, a cross-reference is usually a hyperlink (in blue text).

CV (Curriculum Vitae): A short one-page résumé.

Deadline: The latest date that a piece of assigned writing is due on for submission.

Decision makers: Persons in an organization that determine future directions for the organization; frequently called *upper management*.

Deliverable: See Project deliverable.

Denotation: The exact meaning of a word, without the feelings or suggestions that the word may imply.

Developmental research: Constitutive of studies that are intended to inform development activities practices.

Diction: A writer's choice of words, phrases, sentence structures, and figurative language, which combine to help create meaning.

Disciplinary culture: Modes of work typically through team or individual projects, normative practices, and procedures that encourage consultation or make work processes as well as products transparent; amount of oversight due to presence of external funding or human subjects review; attention to ethics in curriculum.

Dissertation: A thesis written by a candidate for the degree of Doctor of Philosophy; a dissertation may be defined more broadly as any formal, written essay, or treatise on a specific subject; see Thesis.

Document management: The management of documents from their creation to their final disposal (destruction or archiving) in an organization.

Document plan: A step-by-step plan process for how to prepare a document; for example: needs analysis and specific work plan.

Double entendre: A phrase that can be interpreted in two different ways.

Draft: A completed version of a writing that may be rewritten, revised, or polished.

Dummy: A hand-drawn mock-up of what a page will look like in print.

Echo words: Words that remind readers of information they have already encountered.

Edit: To review a piece of writing to correct grammatical, spelling, or factual errors; often includes shortening or lengthening of a piece of writing to fit an available space before publication.

Editor: A professional commissioned to edit (and sometimes write) articles for a publication.

Editorial: A short article expressing an opinion or point of view; often, but not always, written by a member of the publication staff.

Electronic submission: Submission made through electronic means—that is, e-mail or computer disks.

Engineering paper: A published report of the results of original engineering research.

Engineering writing: The kind of writing that communicates the results of research to either nonscientists or nonengineers as well as scientists and engineers.

Ethics: The decisions, choices, and actions (behaviors) we make that reflect and enact our values. A set of standards of conduct that guide decisions and actions based on duties derived from core values. The morality of our actions, study of ethical theory, and the philosophical reflection on morality's nature and function. Also, a set of moral principles resulting from reflection on morality. Synonyms: ethical theory, moral philosophy, moral theology, or philosophical ethics.

Executive summary: A summary at the beginning of a report that gives only the most important information to upper management; not usually targeted for scientists or engineers; sometimes incorrectly called an Abstract; see Abstract.

E-zine: Electronic magazine; a magazine published online.

Fair use: Reproduction of short excerpts from a copyrighted work for educational or review purposes with suitable acknowledgment of the source; usually does not infringe upon the writer's copyrights.

Fees: Money paid to the writer for his or her services.

Figure: See Graphic.

Figures of speech: Ways of using language that deviate from the literal meanings of words in order to suggest additional meanings or effects; not usually recommended for scientific and engineering writing.

First electronic rights: The rights to publish a piece of writing electronically (online) for the first time; once the rights have been assigned, the work cannot be published in another electronic medium; however, reprint rights can be sold.

First print rights: The rights anywhere in the world to a piece of writing in the medium it is published in.

Flat fee: Money paid to the author for his or her work in one lump sum—the author does not receive any royalty after this payment.

Font: An instance of a typeface; for example, "Times New Roman 12-point Bold."

Footer: Information that is repeated at the bottom of each page; see Header.

Forecasting statement: Statement that states the organization of what lies ahead, often appearing with a topic statement.

Formal classification: Grouping items according to observable characteristics that every item possesses; see Informal classification.

Formal report: A report that has a cover page along with front matter and back matter.

Format: The way a document or presentation is arranged. Format includes such things as references in a document and time limits in a presentation. In engineering, there is no single ordained format. Whatever the company, whatever the journal, the writer must arrange the document or presentation to meet the prescribed format of the company or journal.

Formatting: The manner in which a manuscript is prepared and presented.

Front matter: Information that comes before the primary text; features of a communication that precedes the opening chapter or section such as the title page, executive summary, or table of contents.

Galleys (galley proofs): The initial typeset of a manuscript sent to the author for review before it is printed—type size and column format are set, but page divisions are not made.

Glossary: An alphabetic list of terms in which each term is explained; a small dictionary that helps readers to understand the meaning of the terms in a document; often recommended for inclusion in books and other scientific and engineering publications.

Grammar: The rules that govern the structure of a particular language; the system of inflections and syntax of a language; the branch of linguistics that involves the study of the rules of language.

Graph: A pictorial representation of quantitative data that is useful for showing trends.

Graphic: An item of art, for example, a diagram, a flow chart, or a photograph.

Guidelines: A document or statement that defines the standards, the expected actions, or the instructions for submitting work to a publication.

Header: Information that is repeated at the top of each page; for example, headers in printed documentation show the chapter title and page number; see Footer.

Headings: *Signposts* in a communication that inform readers of the subject matter of the text.

Higher education: The realm of education that should take an individual's level of development, in terms of knowledge, attitude, and skills, to a more refined or higher level.

Hook: A type or style of writing used in a title, abstract, or lead paragraph of a work that gets the attention of the reader and keeps him/her reading.

Hyperbole: Deliberate exaggeration; not welcome in scientific or engineering writing.

Idiom: A group of words that has a different meaning from the usual meaning of the separate words. For example, "raining cats and dogs" is an idiom that means "raining very heavily."

Illustration: The meshing of words with images. There are two types of illustrations: tables and figures. Tables are rows and columns of words and numbers. Figures are everything else: graphs, photographs, drawings, and diagrams.

Imprint: A division within a publishing house that deals with a specific category of books or journals.

Informal classification: Grouping items together when there is not a consistent principle of classification or when there is overlap between the categories; see Formal classification.

Jargon: Technical language that is used in a profession, or by a group of people; suitable in some documents but not good for the average user; specialized language/terminology used by a group or profession; language that is only understood by a selected group of people.

Kerning: The amount of space within the text—depends upon the font used.

Keyword: A word or phrase that is related to the contents of the scientific or engineering paper; users can search for keywords, and if that keyword exists, one or more topics are shown to the user.

Kill Fee: A compensatory payment made for an assigned article that was completed but not used or published.

Language: The way we use words. Language is more than just vocabulary; it includes the order of words, the lengths of sentences, and the use of examples.

Language edit: The process of checking for language errors (generally grammar and spelling).

Letter spacing: Generally used in typesetting—the amount of space between the letters of text.

Line length: The width of text or distance of the line.

Line spacing: Space between two lines of text.

Literature cited: A section of a scientific paper in which the citations (references) are listed; typically, this section follows the text; see Reference section.

Manuscript: The final draft being prepared for the printer.

Margins, page: Space around the text on the top, bottom, left, and right of a page.

Mechanics: Usage, grammar, punctuation, and spelling.

Memo: Short for *memorandum*, a brief written communication that follows a format specific to the company in which it was written; memos are written from someone within an organization to others inside the organization.

Minimalism: An action-based and task-oriented strategy to documentation; gives emphasis to what readers must do. A minimalist document contains only information that is important. The basic principle is task-orientation. Brevity is important because brevity can help task-oriented activity; does *not* include creating thin (small) documents by excluding important information.

Myth: A story that attempts to explain events in nature by referring to supernatural causes, like classical deities heroes; should not be used as the basis for hypotheses in scientific and engineering writing.

Name-year system: A system of literature citation in which references are arranged alphabetically in the *Literature Cited* section but cited by name of the first author and the year in the text.

Narrative text: Text that tells a story or describes an event; unusual in scientific and engineering manuscripts.

Numbering, figures: The rules by which figures shall be numbered within the document; this could be continuous at the level of document (Fig. 1) (Figure 1) or continuous at the level of chapter (Fig. 1.1) (Figure 1.1), etc.

Numbering, headings: The rules by which headings shall be numbered within the document; this could be Alpha (Chapter A) or Numeric (Chapter 1) or a combination of both at the sub chapter level (A.a) (A.1) (1.a) (1.1), etc.

Numbering, tables: The rules by which tables shall be numbered within the document; this could be continuous at the level of document or continuous at the level of chapter.

On acceptance: The time when payment is given to the writer after the editor accepts the finished nonfiction article.

On publication: The time when payment is given to the writer when an article is published.

On spec: When the editor is not obligated to publish the piece as the writer was not officially assigned to write it.

Outline: A brief description of the main points or sections of a written document that make it more navigable and organized.

Page design: The rules for setting up a page that have been adhered to in the creation of the document.

Parallelism: Arranging sentences and lists with similarly constructed words and phrases.

Partitioning: Dividing an object into separate parts in order to describe the object.

PDF: See Portable document format.

Peer group: A group of scientists or engineers, generally working in the same area or with the same interests or responsibilities and at about the same level, used to informally review the work of another of the group.

Permission: A fee paid by anyone who wants to reprint part of a book for uses such as excerpts of the book appearing in an anthology; another writer using more than fifty words from the book in a published article. The publisher handles permissions for the author, and splits the proceeds.

Phantom readers: People who will read a document even though they were not the intended audience.

Plagiarism: Presenting the works of another author as one's own work.

Portable document format (PDF): The preferred form in which a document should be exchanged online; this format was created by Adobe systems, and it is very transportable across different computer platforms.

Preferred term: A term that is used in preference to another equivalent term.

Premise: The question or problem that is the basic idea of a scientific or engineering paper, report, or article.

Presentation: The slide content and notes needed to visually and orally communicate an idea or hypothesis to an audience; the slides may contain graphs, bullets, pictures, and/or paragraphs to communicate the desired information.

Print on demand (POD): Publishing a book or books or journals as they are demanded by the potential readership.

Procedure: A list of steps that a user does or needs to complete a task. Sometimes it is suitable to include information about results of a user's action, but usually, descriptive text is best avoided.

Professional: An individual who prioritizes quality service to customers and all that would enhance it, paying attention to the highest ethical standards.

Professional writing: Writing that takes place in the workplace that is persuasive, legally binding, and may address complex audiences.

Professor: One who is an authority and provides leadership for the further development of a discipline, department, or faculty.

Project deliverable: Any output that has been produced by a project team that is expected or promised. Examples of deliverables are project plan, business case, post-implementation review, logical design, database design, implementation plan, test plan, and completed software applications; deliverables are identified in the project management plan.

Proof reading: A stage in the technical writing process where the final draft is read by a proof reader for error; the a final act by an author before publication; includes checking the correctness of headings, subheadings and sub-subheadings, table of contents, indexing, footnotes, captions, referencing, cross-referencing, bibliography of sources, spelling and the proper use of terms, glossary of terms and acronyms, list of figures, list of tables, list of scientific terms and equations, list of organizational rules and regulations, list of cases, list of statutes, list of laws and regulations, list of constitutional provisions, etc.

Proposal: A document that is supposed to persuade the potential funding agency to support a project; the written plan or response put together for the consideration of another person or company; typically includes an overview of the requestor and their situation, the scope of the proposed solution, the resources required, the approach or methodology to be used, a timeline or schedule, and the associated pricing for the solution.

Pseudonym: An alias used by a writer desiring not to use his or her real name; also known as *pen name* (*nom de plume*); not recommended and often rejected for scientific and engineering writing.

Public domain: Any material that can be freely used by the public, and does not come under the protection of a copyright, trademark, or patent; usually works published by the federal government.

Pure research: See Basic research.

Query: A one-page letter to an editor requesting information about a submitted manuscript; also a similar letter to an editor suggesting an idea for an article or a book—usually consists of an introduction, some background on the topic, and a synopsis of the writer's credits.

Reader-centered approach: Writing that considers readers' situations, goals, and expectations.

Record of submission: A formalized record of where and when an author has sent article or manuscript submissions.

Referee: A person who reviews a scientific paper to determine whether it should be published; see Reviewer.

Reference: A document (article, book, etc.) containing pertinent information about previous research that serves as support or proof; a formal entry directing the reader to previous research that serves as proof or support of a hypothesis.

Reference section: A section of a scientific paper in which the references (citations) in the text are listed; see literature cited.

Reprints: Copies of previously published articles made available by the publisher for the writer's files and records.

Résumé: A document containing a summary of a scientist or engineer's education, professional experience, and job qualifications.

Reviewer: A person who conducts a critical evaluation of a scientific paper with the primary purpose of determining whether it is suitable for publication; see Referee.

Revise: To study again; to improve, correct, amend.

Revising: Making changes to improve the writing.

Revision: A version of a document that has been improved, corrected, or amended; the result of revising; the act of revising.

RFP: Request for proposal.

Rights: Ownership of all the various ways in which a creative work may be reproduced, used, or applied.

Rough draft: The first organized version of a document or other work.

Royalties: A percentage of the cover price of a book paid to the author. Royalties are only paid after the book has earned revenues and are usually paid on a six-monthly or annual basis.

Running foot: See Footer.

Running head: See Header.

Science writing: The kind of writing that communicates the results of research to either scientists or nonscientists (cf. scientific writing).

Scientific paper: A published report of the results of original scientific research.

Segmenting: Dividing a writing process into separate parts in order to describe the process.

Serif: A serif is a small projection on a character; sans serif typeface is a typeface that does not have projections:

Simultaneous submission: To send a submission to more than one journal at one time; generally an unacceptable practice.

Solicited manuscript: A manuscript that an editor has asked to see.

Structure: The strategy of writing. Structure includes both the organization of details and the emphasis of details.

Style: The way a writer presents information in a document or presentation; includes such things as the way the writer organizes details, the words, and the illustrations; conventions with respect to spelling, punctuation, capitalization, typographic arrangement, and display.

Style guide: Rules and guidelines that tell authors how to write documents. A style guide usually contains information about the sentence style, layout, typeface, captions, headers, footers, and other parts of a document.

Submission guidelines: The guidelines given by the editor or the publisher for submitting queries or completed manuscripts to the publication.

Summary: A short description of the main points of a body of a scientific or engineering work.

Synopsis: A brief summary of a manuscript or book, written in present tense prose, which is usually two to three paragraphs in length.

Table: A presentation of data in which alphabetic or numeric values are arranged in rows and columns.

Tear sheet: A sample of an author's published work; the actual torn page containing the article.

Technical communication: The communication of a technical message, often in nontechnical terms; sometimes, the term is used as an alternative for technical writing; not applicable to *scientific publication* or *engineering publication*.

Technical writing: Writing that conveys information that is difficult to understand in a clear, concise, correct, and compelling manner; see also Technical communication.

Terms: The agreement made between the writer and the editor/publisher for the publication of a particular work—including types of rights purchased, payment schedule, expected date of publication, and other similar items.

Text: Any of the various forms in which a writing exists; the main body of a manuscript, as distinguished from the title, appendix, footnotes, etc.; type, as distinguished from illustrations, margins, etc.

Theft: Presenting the idea(s) of another person (published or unpublished) as one's own idea(s).

Thesis: An original dissertation on a specific subject, especially one written by a candidate for a diploma or degree; the purpose of a thesis is to demonstrate that the candidate is capable of original research; see Dissertation.

Tone: Whatever in the writing shows the attitude that the writer has toward the subject or the audience.

Topic statement: A statement that increases reader usability by explicitly stating what a paragraph is about.

Trade article: An article written for a specific business trade or industry and which contains information that is relevant to that industry.

Trade magazine (professional magazine): A magazine published for a specific industry or type of trade; typically contains advertising content and job availability focused on the industry in question with little if any general-audience advertising.

Transitions: Writing that allows the reader to understand how adjacent parts of a communication are connected.

Typeface: The name of the design for a set of fonts; for example, "Times New Roman" or "Tahoma." The terms font and typeface are not synonymous.

Typographic contrast: Use of different sizes and weights of fonts to create a distinct difference between elements; best used to create an effective difference between headings and body text.

Unsolicited manuscripts: An article or book manuscript that a publisher did not request.

Upper management: See Decision makers.

Usability: The ease with which readers can understand a written communication to perform their specific task.

Usage: The way in which words and phrases are actually used within a community; writing practice that is accepted generally.

User guide: A document written by a technical writer to give assistance to people using the system.

User manual: User guide.

Voice: The style, tone, and method of writing with which an author composes a work.

White paper: A report or document used to educate, identify problems/issues, and provide possible solutions; also used as a sales and marketing tool, as a means of promoting the benefits of a specific business approach, technology, or product.

Withdrawal letter: A politely worded letter to a publication or publishing house withdrawing a manuscript from consideration.

Word count: The number of words in a manuscript.

Work for hire: A job where the writer is commissioned to write an article but does not receive any author's recognition and does not get any rights to the work.

Writer's block: The inability to write for some period of time. It can be the inability to come up with good ideas to start a scientific or engineering manuscript, or extreme dissatisfaction with all efforts to write.

Writer's guidelines: A set of guidelines that a publication wants the authors to follow.

Index